A NEW LANGUAGE FOR
ENVIRONMENTAL DESIGN

A NEW LANGUAGE FOR ENVIRONMENTAL DESIGN

Lynden Herbert

New York University Press
1972

Contents

Contents

Diagrams

In the Beginning . . .

Societies often reach genuine turning points, periods of time when crucial decisions have to be made which will affect the future for many generations to come. Sometimes wrong choices have been made, or problems approached with bewilderment and indecision that has led to decay or stagnation. Other times, courage and understanding have led to a breakthrough and forward movement.

Such eras of choice are accompanied by a widespread sense of historical loss, by a feeling that existing values and institutions no longer seem adequate, by a sense of disquiet, unease, and nameless dissatisfaction such feelings give rise to a craving for the past or a hope for revolutionary change.

Our age seems to give rise to little enthusiasm. Personal commitment is rare, particularly in the industrialized nations. The discussions are about the growing distance of people from each other, from the social order, from work and pleasure, and from the values and heroes of the past which give structure, meaning and coherence to life.

There is a great confusion about what is valid and good. More and more questions are being asked about what the technologically advanced nations have to offer and the price that has to be paid in return. The prevailing images of Western cultures are of disintegration, decay and despair, not of constructive hope but of awful and imminent destructive potential ready to be unleashed at any time. The images are not of commitment to the future but of alienation from the present.

3

In spite of the gains offered by technology in terms of decreasing suffering and want and of increasing opportunities, there is a grow- ing sense of loss, unrelatedness and lack of connection, and a feeling of alienation that effects even those who go without the least. It is occurring in the strangest form in the post-industrialized societies which are dedicated to mass consumption, mass groups, mass or- ganizations and specialization. These societies are concerned with techniques, the glorification of technical competence, and the sys- tematic application of one branch of knowledge called "science" to the problems of production and consumption. The material Utopia of a century ago has arrived together with the values and concepts of that time, but it has not brought the expected joy with it.

With the age-old goal of universal prosperity in sight, there is increasing question as to whether the values and virtues that made it possible are sufficient to take us beyond the urgent need for goods, money, gadgets, and commodities. The pursuit of quantity, the cult of competition, the rat race have become too high a price to pay for this prosperity. At a time of great affluence there are signs of dis- content, confusion, aimlessness and alienation, which express the bankruptcy of values and visions based entirely on technology and its glorification.

In the next decades, Western societies, and more particularly the United States, will be called upon to decide whether to turn the clock back so as to recreate old societies and outworn ideologies; to continue the unguided "triumphant" march of technology that gives rise to the same alienations and compounds them; or to begin to define new visions of a society where technology is used for truly human purposes.

But there seems to be a fundamental unwillingness, in a period of unprecedented change brought on by technology, to look beyond the process which is responsible for so many problems. Those most concerned about solving the problems created by technology do not seem able to conceive of any course of action except in the same technological terms of quantity, comparison, statistical norms and economic output.

The solutions offered would apply technology to reconstruct worlds that technology has destroyed, to reduce complexity that technology has created, to limit choices and opportunities that technology has made possible. The assumption is that all that is required to solve problems of urban sprawl, disorganization, pollution, congestion, destruction of natural resources and the feelings of alienation, is simply to apply technology on an even larger and more expensive scale. But these processes, values and methods are the creators of these problems, they do not solve them.

The central problems created by an exploding, unplanned technological society merely increase and are magnified by the application of even more technology. New values and methods to exploit them have to be found outside of science and technology; visions of the good life must look beyond material prosperity. The quest for personal fulfilment beyond material goods becomes both urgent and possible. There has to be a commitment to human wholeness, to a capacity for exuberance, passion, laughter, dedication, concern and care for other individuals, and the earth, air and the other living things that mage up the planet on which we live.

There is no doubt about the seriousness of our present problems and the bewilderment and confusion that they are causing. At the same time it is proving extremely difficult to find an explanation for these problems which offers a possible solution that does not at the same time compound the problems. Nevertheless, there does seem to be such a possibility in the idea of "Information Overload." This idea suggests that there is simply too much information around. Any organism, any organization, any system, including those that make up the environment, is so constructed that it can deal only with a limited amount of information. Any amount in excess of this causes the system to become overloaded with information, that is, energy, and to disintegrate, explode. In the last few decades our indiscriminate inquiries and indiscriminate looting of the planet have produced more information than we ourselves, our social organizations or our environment can deal with. We even talk of situations being literally mind-exploding.

Information can be seen as a form of energy that organizes living systems. Such systems maintain themselves by the transfer of information/energy between themselves and their environment. A system is adapted to deal only with a specific and limited amount of incoming information which represents its maximum development potential. This adaptability is expressed by its behaviour which may evolve over long periods of time in response to slow changes in its environment but will eventually reach its maximum development in a highly specialized and inflexible form where only very precise incoming information becomes acceptable and the rest censored out. However, a sudden change in the environment produces a mass of new information that has the effect of overloading it. In this situation only much less specialized systems which have not yet reached their maximum potential have the possibility of survival.

Man, as a species, owes his apparent supremacy to a unique feature that could at the same time prove to be his undoing. Instead of having to adapt his behaviour to his environment he can adapt his environment to his behaviour. His ability to store and pass on information has allowed almost unlimited possibilities of behaviour and by developing electro/mechanical extensions to his body he has suddenly found that he can become bird, fish, giant, dwarf almost at will. In a few decades his evolution has been speeded up to a point that would have taken another species eons of time. But this unique relationship with his environment has also meant that it too has been called upon to speed up its evolution without the normal checks and balances of time. This is clearly proving impossible and the result is an information overload in reverse. We are confronted now with an ecological crisis with the total environment hopelessly disorganized. Pesticides, fertilizers, bad air and water are no less information overload on the total environment than they are on the human organism. Where our minds and its extensions may have speeded up evolution, our bodies suffer from overloaded hearts and lungs and cells. Medical research can be seen to be making desperate efforts to find ways to speed up our bodily evolution.

But this is not everything. All the symptoms of information overload in individuals can now be seen in our society. We pride ourselves in the Western World on the maturity of our organizations and institutions. We have assumed that the environment can always be changed to our ways and have in consequence allowed to develop very unadaptable and overspecialized organizations. The process of censorship of new information can already be observed exactly at the point where we are aware that the environment cannot be treated with disdain indefiniately. If there is any basis for doomsday predictions it clearly does not lie so much in endless circular arguments about the effects of overpopulation, famine, disease, and so on, because these are merely symptoms of information overload at a local level, but in the workings of too much information.

Information overload begins at the point where a system has reached its maximum development. For example, in a social system or organization the first signs appear when an attempt is made to formalize the purpose and procedures so that incoming information can be better selected. It becomes a statement of the area of specialization. This situation could continue almost indefinitely, but new information, seen as change, applies pressure to adapt which can be met only with the equivalent of cries for "law and order." The organizations version of "law and order," with the intention of holding back this change so that the organization itself will not have to change. As outside changes increase, the organization seeks ways to control them without actually affecting its own internal workings. This search for "a magic formula" that has a capacity to make both those inside and outside believe that they have a common purpose could be called "Viet-Namization." At this stage there is a noticeable reluctance for some individual units in the organization to accept these internal policies, deciding at first to do only that which is expected of them, then looking after only themselves, and finally recognizing that they represent the disintegration of individual units, they "drop out," and the organization's collapse is complete. It is a characteristic that potential for integrating an organization only exists before formalization and that it is useless to try to pick up and reorganize the disintegrating pieces because the potential for putting these pieces together again has already run out and any new information will only speed up the overloading effect.

It is not difficult to find examples of this phenomena at different stages of development throughout our present society both amongst individuals and organizations. On the one side we can observe the political system, the economic system, the educational system, the cities, and even the nuclear family trying desperately to deal with information overload and on the other side the beginnings of inquiries into more flexible life styles and organizations such as free universities and schools, multiple living groups, and the liberation of women, men, minority groups and so on from the servitude to overspecialized and inflexible social organizations, so that they can relate much more closely to an environment of their own choosing without overloading it.

The basic argument of this book is that we cannot indefinitely suffer from information overload without jeopordizing our survival; that the price we will have to pay for a fully integrated human life will be a much lower level of want than we know technology can fullfill, and that many of the problems we now have are self-fulfilling, an endless treadmill of endless tasks of our own creation and unnecessary for our happiness and well being. We all live in an environment of our choosing. The environment is the way we see it not the objects in it. If we see it differently it is different. What is not different is each person's need and potential as an individual. We have become so concerned with what other people are doing that we have forgotten to ask ourselves what we are doing. Other people's problems, their needs, have become our problems, our needs. Design is concerned with discovering the most suitable environment for our own or our collective needs. It is concerned with inventing environments by the control of information. Why we design and what we design and how we design is part of our biological survival, part of the living and learning process. The magnitude or even necessity for a product or a construction, is entirely dependent on the environment we need and choose. Present circumstances suggest that we cannot go on choosing the environments that we now have and so the task of us all is to recognize that we are, by our very nature, "designers," and that learning who we are and how we make our own environment, the invention of a "life style" for ourselves, is what we mean by the word "Design."

INFORMATION OVERLOAD

ideal

collapse
drop out

individualisation

magic formula

law and order

maximum development

start of any new organisation

potential or organising ideal

actual development

disintegration

input over time

output

Section I

The Nature of Man

We are determining what kind of organisms we are going to be; Our cities and houses are creating different kinds of people in the slums and the suburbs.

BUT AT THE MOMENT WE HAVE CHOSEN TO IGNORE THE EFFECT THAT THIS IS HAVING ON THE ECOLOGY OF OUR PLANET THAT WE CALL THE NATURAL ENVIRONMENT.

It is known that no species can exist without an environment, or successfully exist in one of its own making; no member of a species can survive except as a non-disruptive participant of an ecological community. Every member must adjust to it and to other members of the community in order to survive. Man cannot be excepted from these biological tests.

The way in which man adjusts is by reading meaning into what other men do in order to establish a balanced relationship of himself to his environment and to his fellow man. This communication and exchange of information is called culture. No man can act or interact at all in any meaningful way except through the medium of culture. It is likewise a mistake to assume that man and his environment are separate and not part and parcel of the one interacting system.

CULTURE IS COMMUNICATION AND COMMUNICA-TION CONCERNS CULTURE.

Spoken language is only one channel of communication of culture. There are others that reinforce or deny what has been said or written in words. The use to which we put time and space, food, learning, play, fighting and the physical extensions of our bodies, are also what we communicate to others. To transfer different habits from one culture to another without general acceptance can be extremely hazardous, whether eating habits, dressing habits or political habits.

So, any approach made to environment problems must be sufficiently adaptable and flexible to encompass this entire environmental matrix and we must see the situation as being in a constant state of flux. We have to untangle the patterns of one cultural environment from another. All options must be left open and all definitions must be made in terms of 'what is possible' rather than 'what must be done.'

TO BE VIABLE WE MUST LOOK TOWARDS PREVENTION RATHER THAN CURE.

1

An Animal from the Past

Out of the dreaming past, with its countless legends of steaming and boiling seas, and gleaming moving glaciers; mountains that heaved and moved upwards; suns that burnt; emerged this creature . . . Man.

The latest phase in a continuous and continuing process that stretches millions of years before him back to the beginnings of life and back beyond even that.

His is the heritage of all that has ever lived; he still carries the vestiges of snout and fangs and claws of species long since vanished; he is the ancestor of all that is to come.

We should not regard him lightly . . . He is you and I.

EVOLUTION

Orthodox theory explains evolution as a series of random tries preserved by selective reinforcement in mental matters, and random mutations, monkeys on typewriters, preserved by natural selection in biological matters. Mutations are defined as spontaneous changes in the molecular structure of the genes, and said to be random in the sense that they have no relation whatsoever with the organisms needs. In this way most mutations are seen as harmful but a few lucky hits are preserved because they happen to give an individual some small advantage. Given enough time there is the possibility that all these little changes will add up to a big one, or a big change will turn up on its own; like an eye or a brain, which will give its owner such a survival advantage that its offspring will prosper. It is never explained how the original species managed to survive adequately without it. The argument is somewhat like saying that if a man stood in a field throwing bricks around for a long enough time he would be able to construct the house of his dreams. There is clearly something missing from this kind of explanation.

The vertebrates' conquest of dry land started with the evolution of reptiles from primitive amphibian forms. These amphibians reproduced in water and their offspring were adapted to live in water as soon as they were born and before they were born. The novelty of the reptiles was that they laid eggs on dry land and no longer depended on water; but the unborn reptile in the egg still needed an aquatic environment. It needed water, food and a container so that the water would not evaporate to get rid of waste products in the shell and some kind of tool to get out; combined with all this was the necessary and countless other essential transformations of structure and behaviour. The changes could be gradual but each step would have to have been harmonious and in the right order. The liquid had to have a shell but there also had to be a tool to break it open. Any other way the tool would have been useless and so would not have survived or the shell unbreakable and the animal trapped unable to get out. Each change taken by itself would have proved harmful and working against survival.

Mutations occuring alone, preserved by natural selection, would not have waited a thousand or a million years for mutation 'b' or 'c.' It would have been wiped out long before it would have combined with the others. All are inter-dependant not independant. They could not have come together due to a blind series of co-incidences.

The alternative is concerned with the concept of hierachies or systems.

NATURAL SYSTEMS OR HIERACHIES

It takes 56 generations of cells to produce a human being out of a single fertilized egg cell. This is done in a series of steps, each of which involves the multiplication of cells by division and their subsequent growth in numbers; and the structural and functional specialisation of the cells into the shaping of the organism. All are complementary aspects of a unitary process arranged in a hierarchic form with rules and strategies.

There is no reason why in this situation the chance mutation of a gene should not affect a whole organ, or a reproductive system, as a whole in a harmonious way if it happens sufficiently high up the system; in the original genes which direct the development of the organisms as a near copy of the parents in all essentials. A change in the hereditary factor could bring about a shift in the whole system and alter a complex organ as a whole as the assemblies at the top of the hierarchy guide the strategy of the development of all those assemblies below that are dependant.

The survival or usefulness of any such mutations would be subject to the rules however. The rules of self-regulating selective powers, as well as the flexible strategies in the growth of systems, on all the levels of development. These controls would eliminate harmful mutation and co-ordinate the effects of the acceptable ones. The screening process begins at the very start at the level of the molecular chemistry in the chromosomes like misprints in a book. However the hierarchy of correctors and proof readers eliminate them if the result is harmful. There is a theory that the ill effects of the drug thalidimide was due to its effect of damaging these censors and so allowing an incorrect foetus to develop. The purpose of these censors or monitors is to prevent instability and satisfy certain stringent tests of physical chemical and functional conditions to survive.

It would seem that long before survival in the external environment is put to the test a series of internal selection tests must be passed for suitability and fitness. A whole hierarchy of internal processes are at work to apply a strict limit on the range of possible mutation, reducing the importance of chance factors.

Instead of proposing that a monkey at a typewriter working over a million years might produce a line of poetry by chance; it can now be seen that it is more reasonable to say that if the typewriter is seen as a machine that is only programmed to print existing syllables in English and to obey all the rules of possible associations of these and the rules of grammar and syntax; and to eliminate anything else as nonsense; then and only then would it be possible to have a line of poetry which would be different from any other.

Evolution of life and life systems is a game played according to unchangeable rules which limit the possibilities but leaves sufficient scope for a limitless number of variations much as in a game of chess. The rules are inherent in the structure of living matter, and the variations derive from the flexible strategies. Life is not a free-for-all, nor is it a pre-determined computer programme. The purpose in every organism from inception is to make the best of its limited opportunities. Purposiveness is a goal-directed instead of a random activity, full of flexibility bounded by rules.

The most fundamental principle of evolutionary strategy, like the composition of language, or the putting together of watches, is the standardization of the sub-assemblies. As in a motor car the various sub-assemblies are self-sufficient enough to be transferable and adaptable to other cars; the engine, the wheels and so on.

Each unit has been developed over a long period and has not changed very significantly since its inception. The wheel of a car does not differ in any essential principle from the earliest wheels. The structure of the forelimbs of a man, a dog, a bird, a whale are of the same basic design of bones, muscles, blood vessels and nerves. Their functions are very different, but as in car manufacture, the basic existing component is modified rather than the invention of a totally new component. The incorporation of a totally new component into car manufacture is further complicated by the fact that there is some plant already producing the component that is to be superceded and this has to be stopped, as well as needing a great deal of modification to that which goes on existing.

In a stable evolutionary pattern, as in the sources of knowledge, it is the traditional that is modified. It is the traditions that are the sources of innovation both in ideas and evolution. There can be a great deal of variation in a limited number of directions but sooner or later these possibilities run out and stagnation follows, the line of development comes to a dead end.

There have been new forms branching out of the evolutionary tree; and the opposite phenomenon in the decline and extinction of other forms. Those that have not perished have become stagnant, their evolution coming to a standstill at various stages; some in the long distance past. The primary cause of this stagnation of extinction is overspecialisation; resulting from an unchanging environment.

There is, however, a possible escape from this inevitability, otherwise new species would never have been possible in the past or the future. This stems from the existence of the phenomenon, in certain circumstances, whereby evolution can appear to retrace its steps along the path that led it to a dead end, and make a fresh start in a new and more promising direction. The appearance of this useful evolutionary novelty is in the larvel or embryonic stage of the ancestor, which normally disappears in the adult stage, but becomes preserved as the adult stage in the descendant. The effect is that the organism begins to breed while still, technically, in the larval or juvenile stage and so expresses all these features in its offspring. The adult stage is never reached and is dropped from the life cycle. This tendency to squeeze out the adulthood and prolong childhood amounts to a rejuvenation and despecialisation of the species, and an escape from an evolutionary cul-de-sac.

RETRACING STEPS TO AVOID DEAD-ENDS

It would seem that this retracing of steps to escape dead ends was repeated at all the major and decisive stages of animal evolution. The conquest of dry land was initiated by amphibians whose ancestry goes back to primitive lung-breathing fish; not the more successful but more specialised gill-breathing fish which have come to a dead end. The human adult approximates to the embryo of an ape not the adult. If the human evolution is to continue along the same lines as in the past it will probable involve a still greater prolongation of childhood and the retardation of the specialised maturity of the adult.

Some of the characteristics of the human adult, for instance, the admiration of rigid and ordered behaviour and thinking and the mechanical skills and demands for specialisation of activity and ways of living; would be expected to give way to more childlike demands for change, flexibility, constant learning and education well into the period of life considered to be adult and mature. At the same time a much lower age of puberty and child rearing when the parents are still in the 'childlike' stage can be expected. It is interesting that this is exactly what seems to be happening in those areas of the world where there has been an evolutionary speed up caused by the acquisition of electrical-mechanical extensions.

The essence of the process is an evolutionary retreat from specialised adult forms of bodily structure and behaviour, to an earlier and more primitive, but also more plastic and less committed stage; followed by a sudden advance in another direction. It is as if there is a backing up the hierarchy where there can be more complex, flexible and less predictable patterns of activity.

The similarity to the present mental and social scene is obvious. The emergence of biological novelties and the creation of mental novelties seem to show certain analogies. In the history of science or art or philosophy, the continuity of ideas only exists in a time of consolidation and elaboration, which follows a major change of view. Sooner or later this consolidation brings increasing rigidity, orthodoxy and heads toward the dead end of unadaptability and over specialisation coupled with a feeling of directionlessness. Eventually there is a crisis, a new breakthrough out of the blind alley and another cycle.

The new theoretical structure which emerges is not built on top of the previous edifice. Instead, it branches out from a point where progress went wrong and by-passes the existing orthodoxy and the institutions that represent it. Their inevitable stagnation, mannerisms and decadence is resolved in a crisis followed by a revolutionary shift in sensibility. Evolution is a history of escapes from blind alleys of over-specialisation and bondage to mental habits and the escape mechanism is to draw back to leap forward again.

BEHAVIOR

We have brains because we think; not we think because we have brains;

The human brain came into existance like eny other evolving

20

structure, in order to deal more effectively with the increasingly demands made on it by increasingly complex behavior demanded for survival. So the first mutant with a bigger brain had greater survival opportunity in terms of natural selection.

The behavioural patterns essential for survival half a million years ago called for an enlargement of the brain with a greater capacity for memory, foresight, self-awareness, conceptual thought. This came about but it also encompasses the previous patterns of the time that went before it. The formative influences of an animal past on the human mentality, and the animal patterns of behaviour which called for a greater brain capacity.

The brain is a development of the nervous system. It proved to be good for survival. If it had had the effect, say, of a third leg that kept tripping us up, then it would not have survived. It proved to be an inaccurate mutation in that it is a great deal bigger than it needed to be to deal with the pressures on the nervous system that new behaviour patterns required; possibly too big by 70% to 80%.

The reason why it's capacity cannot be fully used is because our sensory devices are now too primitive. It is possible to conjecture that in these circumstances, the next mutation that we could expect in the human species would be likely to be an increase in the sensory apparatus and means of perception. In fact it is not necessary to think that things must be so planned. With such a deficiency, and there are probably many others, being recognised by the organism it is reasonable to assume that as soon as a solution to the problem arises, from whatever source, then that will be adopted. And that is exactly what seems to have happened; so quietly that few seem to have realised that anything has occured; not through a change in some genetic organ; but through the development of technology.

The senses have been extended, the means of perception and response increased, so much that there is no way of measuring the future possibilities.

The mistake has been twofold; firstly it was assumed that behaviour followed a new mutation and not the other way around; that it is only organs that evolve for survival value not behaviour that evolved and makes use of anything that might give greater survival value. The second mistake is in the basic assumption that we are separated

21

from the environment, detached from it by out skins. But the environment is ourselves; to increase our attachments, our senses and our perceptions is to extend the environment and bring it under control, so giving greater survival value.

We have invented machines extending the muscles and the time span over which they can be used; new electrical equipment extending senses and perceptions. We can see to the furthest parts of the universe or the smallest parts in it.

To have access to these new extensions, to be able to use them and to understand them is to be a new mutant, a new animal species; not to have access is to be left behind. Not everyone on earth does have, or soon will have, these extensions, nor is it necessary except where they are all freely available. This is one of the basic dilemmas of the so-called underdeveloped countries; that they feel the influence without the possibility of access to them. It is perhaps a hopeful sign that not all wish to run headlong into this situation.

The clash of these traumatic events and the child-like learnt prejudices from parents can easily be felt. "Things are not what they were, everything is going down hill."

The young are more often than not more used to these extensions than their parents and find nothing mysterious or to be frightened by. Cars, television, radios, computers are things easily understood and worked. The new experiences and new information is taken for granted. There is nothing particularly exceptional about electronic pop music; or flying faster than sound; or sending men to the moon for them. They want to explore the next things not call a halt to exploration. They want to take part in the development of what comes next; they want to get on with it rather than wait permission from their elders after they have fitted the new information into old patterns that they find acceptable.

The big under-used brain, now allied to new perceptual organs, makes brute force no longer the greatest strength; muscle and muscle extensions are now giving way to thought; the possibility of the big brain is taking place in response to new behaviour.

Behaviour is the result that is observed when learning deals with experience of the organism's environment.

Although the nature/nurture debate is by no means resolved it does seem more than likely that acquired characteristics cannot be inherited; that within a species every member is born in the essential image of the first of its kind; no child born now can significantly differ from those of earliest man. Man is importantly different in one respect; although man like animals cannot inherit characteristics genetically he can do so by the process, unique to man, benefit from past mistakes by learning from his parents.

It is behaviour in animals as well as human beings that determines whether bodily characteristics have selective advantage or not; it is behaviour that evolves and is central to the evolutionary process.

Behaviour is how we act, it is the only way that the relationship and interaction between animals and their environment can be observed.

BIRDS DO NOT FLY BECAUSE THEY HAVE WINGS . . . THEY HAVE WINGS BECAUSE THEY FLY.

They have wings because their type of behaviour meant that flying gave them a selective advantage; not the other way around. One of them did not suddenly gain wings and survive better than the others; because to acquire such an additional possibility without a particular kind of behaviour would have the opposite effect. Flippers instead of hands would make it easier for us to swim but we could not eat. If there was something intrinsically better in flying for survival value we would all be flying.

It was the acquisition, by chance, of a large brain and the capacity it gave for speech and stored experience that gave man his great advantage for survival and dominance. It gave the possibility of inheriting acquired characteristics; knowledge, the result of previous experience. This we call culture, a series of languages which includes spoken language; and which has to be passed on and learnt in every generation, and superimposed on the genetic inheritance. The cultural record differs from one group to another. The cultural records are responsible for the future behaviour patterns of every human being.

These behaviour patterns are superimposed; on the nervous system very early in life.

23

MAN THE ANIMAL

Man is a vertebrate, a mammal, and a social primate.

The primate group to which man belongs arose originally from a primitive insectivore stock, who hid in the forests while the great reptiles ruled the world. Following their collapse these insect eaters emerged and spread and grew into many shapes. They became plant eaters, and burrowed underground and became clawed and toothed killers, and grew long legs to escape from their enemies; The open ground became a field of hunters and the hunted.

In the forest the insect-eaters broadened their diets to include fruits and nuts and leaves and berries. They began to approximate to primates with eyes coming forward on their heads to give three-dimensional vision, and hands to grasp food. The new vision and the grasping hands needed, and got, an enlargement of the brain; and so their dominance of the trees began.

Some thirty million years ago the pre-monkey acquired a tail and grew bigger, scampering among the trees. Later swinging hand over hand amongst the branches a different group began to lose its tail and set out foraging at ground level. The ape had arrived but lived most of its life in the trees still, munching fruit and not caring to compete with the herbivores and the killers, which had also developed, on the ground.

This situation might have continued indefinitely if there had not been some fifteen million years ago a violent change in climate that began to reduce the forest regions rapidly. The ancestral ape was confronted with two choices; to cling to the little remaining forest which was still diminishing, or to take to the ground. The ancestors of the chimpanzee, gorilla, gibbon, stayed put and has been diminishing ever since. The ancestors of the only other ape, the one without hair on his body, left the forest and threw themselves into competition with the already efficiently adapted ground species, which seemed on the face of it to be totally suicidal, and so it nearly proved.

The prospects in the new environment were bleak but not hopeless. The possibility of becoming better killers than the carnivores or

24

better grazers than the herbivores seemed very small. The digestive system could not deal with excessive grass or meat. But there were areas of possible food that were not being competed for and which approximated to the same diet as the forests provided; vegetables and roots could replace fruit and nuts and berries; and small reptiles and fledglings and eggs and the sick and the young could replace the bugs and insects; and speed and claws were not necessary. The big protein animals were still outside their grasp but the first steps had been taken.

Then suddenly in the last million years things began to happen very fast and altogether, one small change urging on others. The ground apes already had sharp and large brains and grasping hands and good eyes, and a degree of social organisation as all primates; but there was also the strong pressure to increase the prowess in hunting and killing larger and larger prey whose large protein supply would cut out once and for all the incessant need for continuous hunting and eating. So they became more and more upright and better runners. The hands became free of the need to help in locomotion and so came to hold weapons. The brain became more complex and able to make decisions quicker. The way out of the impasse that emerged was not to imitate the specialised killers, but to become something quite different using artificial instead of natural weapons. The hunting ape, the killer ape, was in the making.

The behaviour became more and more complex; the tool using became tool making; weapons improved and so did the social cooperation in hunting; pack hunting developed with a better brain solving the problems of group communication and complex manoeuvers. The brain grew even bigger.

Essentially the males were in the hunting groups; the females stayed behind to rear the young; and as hunting increased in intensity and the forages roamed wider and wider; the hunters had to abandon their nomadic ways and develop a home base for his spoils and where he could be sure that the females were safe in waiting to share the food when it was brought back. The basic foundations of man's behaviour were already being laid down.

The hunting ape had become the territorial ape; his sexual, parental, and social patterns began to be effected. The old wandering

25

nomadic life was fading (but not entirely lost even now). If he went away for a long time he had to be sure that his female would be waiting whenever he returned. Seasonal sexual restrictions disappeared and "pair-bonding" took over so that there was one female each to ensure that all males returned. In consequence her sexual attractions had to increase to make sure that he stayed around her and provided her and her offspring with food during the long childhood period. The leadership of the hunt ensured that the most skilful and the strongest also became the most socially dominant and the hierarchial social organization began to appear to be followed by the opting out of the hunt by the skilled tool makers who stayed at home and in consequence reinforced the necessity of the pair-bond as a piece of social as well as biological taboo. The pressures were now for an even larger brain that could cope with this new demand for communication and taboo learning. So, it came about but the brain in question turned out to be a great deal larger than the perceptual organs could use.

The ape was getting responsibilities of keeping a home with the comforts of fire and food and storage and shelter building and social order. The realms of biology were getting left behind and the realms of culture began to appear. The biological basis of these advanced steps lay in the development of the nervous system in sufficient complexity to deal with the increased pressures on it, but the exact form they assumed was no longer a matter of specific genetic control. The social structure became culturally determined and the restraints had to be learnt rather than biologically modified. Nature could ensure that every hunter was happy when he got home but not that some would stay at home to make tools and weapons and not upset the females while the others were away.

So the forest ape became a ground ape; became a hunting ape; became a territorial ape; and finally a cultural ape. The next stage to dominate the previous stage but none were totally lost even in the final cultural stage. The maker of fire became the maker of space craft in a mere half million years, but beneath the surface there is still an ape and a primate preditor.

A brainy weapon-toting wolf which started as a peaceful fruit-eating forest dweller. The genetic accumulation is still there, the question is where and how much?

The most obvious place to look would seem to be in the closest approximations to himself in the animal world and what has to be looked for is something of the following specification:

—A vertical
—hunting, working, weapon-toting, territory-space defining
—brainy beyond his perceptual apparatus
—overtly-sexed, furless
—primate by ancestry and carnivore by adoption with dominant cultural rather than genetic endowments.

AGGRESSION

The evidence of history, the daily newspapers and television which so constantly depict man's apparent willingness to hurt and kill his fellows and even in some circumstances actually take pride and pleasure in doing it has led to the invocation of the hypothesis that man is anything but naturally peaceful and is in fact dominated by an instinct for aggression and killing. This instinct is thought to be normally under strong restraints in civilised societies but capable of erupting in cases of the breakdown of law and order or in psychopathic individuals or in primitive and savage societies. Such an instinct if ever proved would provide a fine excuse for all the killing and joy in war for which there seems so much evidence.

All known societies make a clear distinction between the killing of their own group members; called "murder" and the killing of outsiders. "Rational" murders for gain or jealousy are understood even if strongly deprecated and punished. However even psychologists are at a loss to explain murder without apparent rational cause like the nine nurses in Chicago or a film star and her friends in Los Angeles. It is implicit in our thinking that there is an inbuilt instinct against killing except under extreme provocation. The term psychopath is only a confession of ignorance of real causes.

27

There seems more than enough evidence, at first sight, that killing must be in the instinctual inheritance of man, but this is not necessarily true. All carnivores have an inate inhibition against killing members of their own species with two known exceptions; man and rats. Animals with potentially lethal teeth or claws or horns can be automatically stopped from pressing an attack on fellow species members by signs of submission or flight. Once the submission signs are made the attacker automatically halts and literally cannot kill the defeated rival. A bomber pilot on the other hand can bomb women and children and return to look after his own all in a days work. Since man is so physically ill equiped to kill it seems that he did not acquire the inhibition against killing other men as part of evolutionary protection against wiping out their own species that the tiger or the wolf did. But then man invented the weapon.

It must be emphasized that for all carnivores, including man, killing other species for food is innately of a different nature from killing members of the same species for rivalry or jealousy or pleasure. In animals there is no connection between hunting, which is non-aggressive, and ferocity towards members of the same species.

Man shows the same aggressive disinterest as other mammals when killing other species for food but when it comes to his own kind the greatest similarity is with the rats where the analogy is almost total. Rats live in packs and do not fight with, much less kill, members of their own pack but are quite ruthless with members of an alien pack. They lack the inhibitions against killing the members of their own species. Humans likewise live in packs, or societies; killing others of the same pack is absolutely forbidden except in very special circumstances and is subject to very severe sanctions. This ban does not apply to members of another pack where home members are expected to reverse their position, and where they find equally strong sanctions if they do not kill them.

To accomplish this, members of the home pack are usually called fully "human" and all others are usually thought of as sub-human, to be called "wogs," "gooks," "charlie," or "krauts," and therefore killing is not murder. This primitive type of rat-thinking is never far below the surface, even amongst the most civilized and gentle. Humans differ from rats in their varying definitions of who should and

28

should not be included in the pack. Usually the definition revolves around speaking the same cultural or spoken language or between whom real or imaginary kinship can be traced.

These tribes are definable as the unit where killing is considered murder, and outside which killing is proof of manhood and bravery; or a pleasure or a duty. If done in groups this is known as "warfare." The nation state managed to extend the murder area big enough to incorporate several tribes, and came to include and protect most of the inhabitants of a common language or geographical area. The nation state was the last successful human invention to extend the size of the pack within which killing is murder.

However, more recently, as numbers increase, the old rat pack ideology seems to be rearing up again. In the human rat-pack, status is denied to all persons who do not share the same hypothetical superiority; whether ancestry, skin color, facism, white supremacy, black power; all justify hatred and contempt for those outside. Recent history has shown how quickly such a situation can develop into humiliation torture and killing.

The evidence is endless that our species has no inhibitions against killing those outside his own pack, however he defines it, but this does not admit to an instinct as part of a hereditary endowment.

Amidst the general slaughter there do seem to be a few primitive and inaccessible tribes who merely ask to be left in peace. They are technologically backward and are unlikely to have had missionaries placed amongst them. The most significant and common trait among them seems to be a great love for living and concrete physical pleasure; eating, drinking, sex, and laughter. There is no confusion of sexual identity, and the model for the growing children is enjoyment and performance rather than symbolic achievements or ordeals to be surmounted or ideals to live up to. There are no heroes and martyrs or cowards or traitors to emulate or despise. A happy, hardworking and productive life to do what each wants is available for all.

There is no reason to suppose that they belong to a different species but it would seem that they differ very much in the values that they have adopted. For them, the value of peace and the absence of quarrelling and jealousy are far more important than a reputation for bravery and virility.

It does seem possible that the development in the present young all over the world for the inarticulate sense of redefinition of concepts of 'man' and 'women' and the repudiation of aggressive masculinity in the previous generation, and which is still maintained by the conventional suggests that a re-orientation of value might be taking place.

If all the present revolutionaries maintain their present scale of values when they instruct their children, then it is just possible that the new societies that they would like to see will come about and the roles of the sexes will change and the glorification of killing will disappear. It is just possible that the youth international with its emphasis on shared sensual pleasure and its repudiation of the ideal of manliness and bravery against a non-existant enemy will succeed where no idealist has succeeded before.

THERE IS NO KILLER INSTINCT ONLY A LACK OF INHIBITIONS.

Aggression follows frustration: cultural frustration which denies the possibilities of achieving the values of that culture or any new values of different evolving new cultures.

2

The Social Animal

While observing the human environment we are always confronted with the dual determinate of it being conditioned by human life and it conditioning man's functions and development. We cannot do anything without permission however much we would like to. This in itself is not enough to provide sufficient guidance for modifying action. On the contrary this knowledge is liable to hide the fact that the incentive to change results from the experience of an incompatibility between the environmental situation, and the human situation; and the motivation is either free human energy or by a deliberate disturbance in the biological or physical equilibrium.

Such conditions urge communities to 'reflect' themselves in new ways on the surrounding biota and lanscape; to introduce new land use and reorganisation of the cultural milieu. But such activity is countered by reciprocal environmental influences which modify all goals and thus a new equilibrium is found.

The question, then, is what can be learnt about man and his social system by the study of animals and animal society? Man is a physiological organisation produced by a growth process which is largely organised by characteristics contained in a genetic code accumulated over a million years, as a result of a very complex mutations and selections.

It is therefore possible that the study of animal behaviour could throw light on human behaviour and perceive patterns that may have been present in his ancestors and is therefore likely to be present even now. However, this can be very dangerous analogy seeking. None of man's direct genetic ancestors are around any longer. Their bones can be studied but not their behaviour. Apes today may have diverged from their ancestors in the last two million years as much as man has from his.

It is a great peculiarity of man that he is genetically endowed by an enormous nervous system of ten billion components, unlike any other animal, and that its informational content is almost wholly determined and derived from the environment . . . that is the inputs from outside. Almost everything that happens after birth is thereafter learned. His genetic endowments limit what he can and cannot do physically but not mentally. Men are indeed animals but they are more than this.

This learning is concerned with behaviour between members of the species and is generally called CULTURE. This certainly has a biological derivation but is much more complicated than the adaptive advantages to an animal species. If a species disappears so do the individual members; if a culture dies out the members go elsewhere.

Man and his proto-human ancestors have been using tools and weapons for several million years. Several crucial peculiarities of man: speech, upright gait, sexual mores, lack of fur, placement of sexual organs, are adaptive and have evolved in the context of a natural environment modified by the use of tools and weapons; but man did not evolve as a complete human-being and then acquire speech and culture. These were developed while the animal species was developing and this process goes on even now. Other animals adapted to a specialized environment which is relatively stable. Drastic changes of environment usually kills off a species.

On the other hand man's adaptive apparatus includes a technology which changes the environment as it goes along. Man and his environment are always out of step. The extreme and ever increasing speed of the ecological and sociological change wrought by the development of technology causes many customs to become mal-adaptive in less than a generation.

Students of animal behaviour are so used to finding a perfect fit between an animal species and its environment that they make the mistake in assuming that there ought to be a similar fit between man and his and look on the lack of one as something pathologically wrong. But the development of speech in homo-sapiens has completely altered his nature.

Animals can communicate a certain number of messages to other members of their own species by gestures that elicit predictable responses; but in contrast, human beings can say an infinite number of things in an infinite number of ways and the responses are by no means predictable. It is true that as a member of a common species there is a pre-disposition to behave in fixed ways that reflect the bio-chemical constitutions. All have the natural capacity for speech and spoken language and how this is formed to convey messages. But what these messages say is unlimited. There are also physical limits to gestures but not to what the messages say. Human customs are not pre-determined in the same way as those of other animals.

This is the characteristic that differentiates cultural evolution from biological evolution. Cultural evolution in the last ten thousand years has come to dominate biological evolution, not through the genes, although this may sometime become possible if a very extravagant way to achieve something that is already being done with other means, but by transmitting information through the learning process and not through the information coded in the genes.

The critical question is the extent to which the genetic composition of the human body limits and determines what it can learn from its environment. There are such limits but they are very far out and much more likely to be a failure in the experiential learning process than pysiological in origin.

The problem is complicated by the fact that environments themselves can breed. Culture is a body of coded information which is passed on from generation to generation, suffering mutation and selection, just as the coded information in the gene is passed on, except that cultural evolution is much more subject to mutation and proceeds at a much more rapid rate than the evolution of the genetic structure.

One of the special peculiarities of the species homo-sapiens is its lack of ecological specialisation, no other species, with the possible exception of the rat, can adjust so readily to drastic variations of temperature, humidity and diet. This highly advantageous plasticity is tied to the fact that the species is almost entirely free from behavioral limitations and attitude to his environment. He is much more likely to change the environment than himself if there is a clash of behavioural requirement. If on occasion man exhibits symptoms of behaviour resembling that of animals, it is most probably because of similar physical limitations that were once forced on him by adaptive behaviour.

HUMAN BEHAVIOUR IS A CULTURAL, NOT A SPECIES CHARACTERISTIC.

Culture is the communication of messages between its members. Messages convey two different kinds of information:

THE STRUCTURE

THE FUNCTION

Structural information in culture is concerned with connections and boundaries available for subsistance-work-hunting, defense-murder, possible mates, and available body extentions. The definition and use of space.

Functional information is concerned with social interaction and association and personal identity.

3

Social Interaction

The term 'human environment' can be used to describe the space that surrounds human movement and activities; the environment as a set of biological and physical facts in and modified by man; the 'real' situation. Man as a powerful agent of change in the environment, himself undergoes change in time. There is something unique in his species, and gives him a role in the natural community of animals for a dynamic and often unstable contradicting relationship with the space around himself. There is a give and take relationship, a dynamic in space, coupled with a recurrent change, destruction, and renewal; a dynamic in time.

All established settlements show evidence for this by the continuous imposition of new patterns of living on the old; a needed temporal situation unaccounted for in the present attitude to planning and designing new buildings and new settlements.

TERRITORY

Everything has a physical boundary that separates it from its environment. In a simple organism the boundary very obviously begins and ends at the skin or other protective covering. But as the development ladder is climbed another non-physical boundary appears that exist further out than the physical one. The new boundary is difficult to delimit but it is just as real. This is called the organism's

TERRITORY. It is a species characteristic and genetically programmed.

'Territory' is behaviour by which an organism, but not all organisms, characteristically lays claim to an area and defends it against members of its own species. It is a group rather than a species property expressing the effects of individuals with one another and the environment. A territory can be static or moving.

'Territoriality' ensures the propagation of the species by regulating density of population to food supply; by providing a framework to learn, to play, and to hide; it co-ordinates a group, provides suitable mates, and holds it together; it keeps members of the group within communication distance so that food or enemies can be signalled or protection given by numbers: It offers protection from preditors; it exposes the weak who cannot defend themselves or a territory therefore ensuring the survival of the species; it also protects the environment from over-exploitation; it promotes dominance and the so called pecking-order. Competition between males is usually for territory and not for females; male birds have the finest plumage; females fight other females for males-with-territory.

Basic to territoriality and to many other uses of space in the animal world is the sharp sense of the limits that mark the distance to be maintained between individuals. Two of these are flight distance and critical distance and are used when individuals of different species meet; two others are when members of the same species interact, and these are called personal distance and social distance.

Flight distance is a basic reaction to survival allowing sufficient time to escape from a preditor.

Critical distance is the zone between flight and attack and can be measured in centimeters.

Some species avoid personal contact, some seem to enjoy it. There is no rule for either.

Those species that avoid it have regular spacings between themselves that might be an internalised version of territory. Dominate animals need more distance which helps control aggression. Man starts as a contact species early in life then avoids contact publically a fact as we shall see later has a very strong influence on his psychological make up.

Social distance is the limit beyond which the animal feels threatened and is psychological.

SPACE

All animals have a minimum space requirement without which survival is impossible; this is the critical SPACE of the organism. When a population has built up far enough and this space is no longer available, then a critical situation arises. The simplest way of removing the tension is to remove some of the excess number. There is clearly some relation between space available and reproduction and population control. This seems to call for a reconsideration of the Malthusian doctrine that ties population to food supply; migration of lemmings and other rodents, for instance, following large scale population build-ups when food is still plentiful.

The relationship between preditor and prey is one of subtle symbiosis in which the preditor does not control the population but exerts a constant pressure that acts to improve the species. It is now recognised that body chemistry working through internal and external body secretions helps to control population and integrate behaviour, and its increase and decrease. Animals can die from shock if repeatedly stressed from overcrowding. Other medical problems similar to cancer and heart attack also result.

These processes of selection that control evolution favour the dominate individuals in any group; not only are they under less stress but can stand more stress. When increased stress leads to increased aggression there is a demand for more space and a chain reaction is set up.

Rats have very many similarities with man in behavioural make-up and it has been found that rats like humans cannot tolerate disorder for long and need time to be alone. If harrassed or overcrowded disruption results in important social functions, and overstress leads to population collapse and death from heart attack and cancer and shock, after complete apathy has set in. The most able to adapt to these stresses survive and become dominant. This situation is known as a "behavioral sink' and approximates to human slums with their signs of aggression following frustration, sexual abnormalities and unexplained killing and robberies as well as general apathy in some sections.

It is evident that man cannot escape the fact that he is an animal and that the use of space is similar to other species. But self domestication has emphasised the biological near elimination of flight reaction and the development of sensory and thermal screens as in the almost complete disappearance of smell and taste in some cultures that allows a greater number of people to be packed into a smaller space. Animals with a good sense of smell can easily recognise the smell of anger and other emotions that make close living an impossibility. Proper screening with walls and deodorants can lessen stress from overcrowding but humans like rats need to get away from each other sometimes and this sets the final limits of packing humans into the smallest areas.

Man, too, has genetically inherited a form of territoriality; the attachment of a man to 'his' chair, or a woman to 'her' kitchen, or a student to 'his' or 'her' place in a classroom indicated that the mechanism for territory still exists even if its uses are culturally determined.

The development of spoken language as an information transmission system has obscured the fact that there are other languages that transmit messages from one individual to others concerning an individual's place in the environment and one of these is the use of territory or in its culturally determined form 'space'.

Culture in all its details is the way space is perceived and used. It is the way that space is coded so that others can know how close to approach, when to advance and retreat, how close to come and so on.

Man's sense of space and distance is not static and has very little to do with the single point perspective, linear in form, so beloved of the Renaissance and schools of art, architecture and planning. Man's perception of space is dynamic because it is related to action; what can be done in a given space; it is a spatio- temporal experience rather than simple passive viewing.

DISTANCE

Man should be seen as a series of expanding and contracting fields or bubbles from which he receives information of many kinds. The decisive factor in distance at any one time is the way people are feeling towards each other; people who are angry will move closer and start shouting as an aggressive gesture; where a man feeling amorous will also move closer but will drop his voice as a means of appeasement and intimacy; the opposite effect of anger. A woman will signal her disapproval of the suggestion by moving away.

There are three main distances in man which coincide with his hierarchic form:

The INDIVIDUAL, or ISOLATE, is in a situation that is associated with the need for expressions of intimacy and physical contact transactions; touching, lovemaking, protecting. Distance: touch to 18 inches.

The PERSONAL, GROUP, or SET, situation of providing a protective sphere from other members of a non-contact species; just holding and touching, friends and near acquaintances, parents, parties. Distance: 18 inches to four feet.

The SOCIAL PUBLIC, PATTERN, situation beyond the limit of domination and personal involvement; without touching; impersonal business and public gatherings at meetings and in the street. Distance: four feet to limit of sight.

4

Association

A elucidation of human evolution would provide a key to understanding the changing shape of the human environment. Some distinction has to be made between the cultural and biological aspects. Whereas biological changes are hereditary, cultural acquisitions can and are altered by each succeeding culture and in each cultural area and it is very necessary to distinguish which process might be under consideration at any one time.

It is possible to gain insights into both, however, by appreciating and noting their inter-connections wherever they appear. In environmental change, cultural changes are in fact incorporated into biotic realities, while biological conditions mould human culture in their time.

In his books E. K. Hall discusses the connection between cultures and their use of space. By doing this he provides a much more viable definition of the word 'space' than is usually assumed in environmental designers' discussions.

Culture is communication and communication culture; culture is learned and shared behaviour. Culture is communication expressed through territory and space or observable behaviour, and is biological in origin and learned through genetic inheritance. Space is the 'structural' function of the cultural message system.

Communication is culture expressed through a number of languages of which spoken language is only one and these are concerned with teaching and learning of the particular rules and customs of a cultural group. The messages are non-biological in origin and refer to acquired characteristics that are not genetically transferable; they constitute the 'functional' part of the message system.

Culture is not one thing but many; there is no basic unit although some seem more important than others in any given particular context. The bases for discussing the languages of culture are deeply rooted in the biological past but what they say is related to a particular group at a particular time.

LANGUAGES OF CULTURE

There is the language of: Interaction; with the environment of which speech and the use of space and time are the most important.

Association; concerned with social ordering, dominance and societies, institutions and the organisation of their components.

Subsistance; concerned with finding of food and work.

Sexuality; concerned with sexual attitudes and behaviours.

Territoriality; concerned with the possession and use of a space to move and work in which is defendable and personal.

Defense; concerned with protection, including fighting religion, medicine, and law enforcement.

Learning; concerned with the acquisition of culture from any source and its symbolic storage.

Extensions; concerned with the exploitation of the environment and the adaption of the body, or tools, to meet specialised demands in the environment.

Culture is concerned with messages and their meaning as functional signifying systems.

CULTURAL INDETERMINCY

There is a principal of indetermincy in culture as a hierarchial system. The more precise the investigation on one level the less on others; as in hierarchies investigation into any isolate in detail immediately turns them into sets in their own pyramid. Only one level can be described at one time.

There is also a principal of relativity operating in culture as everywhere else. Experience is something that man projects on the outside world in a culturally determined form. Man alters this experience by living and by being aware. There is no experience that is independent of culture against which culture can be measured; no authority and no absolute standards.

What is good by some standards may be bad by some other. What is good for one group may be bad or of indifferent effect for another. Good may become in time bad, even for the same group.

It is important to remember that culture is not one thing but a series of complex activities inter-related in many ways; activities whose origins lie in the distant past when there were no cultures and no men. The development of spoken language and its offshoot technology, was made possible by the ability to store knowledge information. By the time man had emerged on the scene a great deal of the evolution basic to culture had already taken place in the very systems that are thought of as basically human.

The use of space is very much a product of the biological past but what it is used for is very much the result of current, and in absolute terms, arbitrary usage; but that usage satisfies individual emotional needs that are culturally expressed but in their turn reflect the biological past. The 'functional' information in a cultural message system is concerned with social intercourse that responds to demands for satisfying personal individual needs for intimacy or personal identity; spontaneity and awareness; and security and stimulation.

5

Personal Identity

If we decay faster than we can reconstitute ourselves then we die. Sensory deprivation is deprivation of incoming information. This is the touchstone of personal identity. We change as we live, we are not stuff that abides unchanged, but patterns that perpetuate themselves, dependant on processing incoming information and reacting in our behavior to it.

SENSORY DEPRIVATION

Dr. Eric Berne lays out in his books a general theory of social intercourse that can be summarized as follows. It gives an explanation as to why people communicate together and why they need to do so. It does not describe so much how people communicate to each other through the use of space, but why they communicate. The advantage of his theories is that it provides psychological explanations without resorting to specialised jargon and unlikely axioms.

It has been observed that infants deprived of handling, over long periods of time, will tend at length to sink into a irreversible decline; if deprived of emotional and sensory stimulation from close physical contact the outcome can be fatal.

For the same reasons internal, emotional, and external, sensory, deprivation in adults can lead to apathy and death.

In effect, the lack of touch and handling can cause a deteriorating in the nervous system. The cause of the dilemma is the change over from a physical contact species in early childhood to a non-contact species in later childhood and adulthood. After a period in the womb and after birth in very close proximity to the mother the infant is then confronted for the rest of its life with the choice of getting back to the original state of intimacy in the infant style and the social and biological forces that stand in the way of its attainment.

INTIMACY

In most circumstances an individual will learn more symbolic forms of handling to satisfy the craving for intimacy, even in the recognition of his presence in various ways which can serve the purpose for a time. This process of compromise between the desire and the amount that can be attained in the public areas that most time is spent, and which is sometimes called sublimation, can be described as a hunger for recognition in the same way as there is a hunger for food. As the complexities of compromise increase so the sign of recognition become more individual and more various. Some actors may need symbolic handling almost continuously while others, like scientists, may be satisfied with the Nobel Prize and a few encouraging words from someone he respects. Conversation denotes the recognition of anothers presence and has an analogy with handling or touching intimately; the intensity and content of the conversations indicating how near two people are prepared to allow each other in intimate contact. Conversation is a symbolic representation of physical touching and constitutes social intercourse and its meaning and purpose.

The principal that emerges is that any social intercourse is better than none, biologically, and that intimate intercourse is the best of all but cannot be sustained socially for very long periods, as there is little opportunity for close intimacy in public life.

STRUCTURING TIME

Time is structured, when physical contact proves impossible, in two basic ways. One involves some sort of mental or physical task or project that absorbs the maximum of concentration and so involves the least amount of social intercourse, to get away from the effort needed; and the other way is in the opposite direction, to attain and satisfy the needs for intimacy but in such a way that it will not be damaging to the seeker because of rejection, or humiliation, at any other the stages towards its attainment.

The function of all social living is to lend mutual assistance to solving problems of unstructured time. The need for stimulation and recognition expresses the need to avoid sensory and emotional starvation, both of which lead to biological deterioration. The need is to avoid boredom, the evil of unstructured time, which has the equivalence of emotional starvation. No man can stay in a state of boredom for long without ill effects as brain-washing techniques have discovered and exploited.

There is no rule as to the amount of each that a person may need or desire and depends very much on original programming and moods of the moment and can be exaggerated to extremes of either. It is possible to create a project or task that will fillup as much time as possible either by physical tasks or mental effort, leaving little time for social intercourse. This situation is most noticeable when the person involved is either incapable or deliberately avoiding intimacy from fright or disinterest. 'Work' is the most recognisable form for structuring time in this way but it is not always adequate. When the work is boring, repetitious, or uninteresting then efforts are made to make social contact among fellow workers. Furthermore as working hours get shorter then filling up 'leisure' time becomes a problem that has to be filled with specially dreamed up activities. A housewife is particularily adept at filling up her day with tasks that could take a much shorter time if there were other things to do, and this happens as soon as she takes a job or finds some other interest.

Solitary individuals structure their time either with projects or with fantasy. In societies where work or small-time household tasks are the only really accepted time structuring events, and where social contact is rare then there is an increasing tendency to adopt fantasy as a time consuming activity. This is found particularily in the watching of television in the evenings where long hours can be whiled away in involvement of almost total fantasy, and complete social dis-engagement.

Work is not a substitute for intimacy, it merely avoids it. Therefore over a long period of work and no social intercourse or intimacy the individual is liable to deteriorate into apathy and a state of near total sensory deprivation, saved from utter fatality by occassional casual and faint recognition by passersby.

Because there is so little opportunity for intimacy in daily public life, and work is no substitute, the bulk of the time is taken up with preliminary engagements as a prelude to engagements in greater depth, in all social intercourse. The preliminary rituals of handshakes, and comments on the weather, and the pleasantries of good manners soon gives way to more serious forms of interest if there seems to be a possibility of more intense personal contact.

The sign, though they usually occurr unconsciously, unless being deliberately exploited, are the transfer from social-public space to personal space. The voice and gestures change and the distances between people are reduced so that occasional actual handling can take place and not just symbolic handling. Conversation becomes a series of trial baloons that establishes the speaker in a role that is thought to be compatable with the listeners, as well as inquiry into the possibility of further stages of intimacy. These kinds of ways of passing time are mainly for the purposes, however, of mutual back-scratching or handling, and have no ulterior motive apart from sounding out vague possibilities. Their most common form is usually a party, or cocktail party where the discussions are of mutual friends, cars, clothes, general or specialised knowledge and so on. Therefore their handling potential is limited.

LIFE GAMES, ROLE PLAYING

The next stage is the transfer from personal to intimate space, or individual space in the first instance. The possibilities of direct personal rejection and humiliation are now very much stronger, they cannot so easily be shrugged off onto someone else, so the stakes become higher and so do the rewards. These moves to more complex and personal relationships require more individual programming and thought and also more clear cut rules. The rules are needed because these games involve not just contest, as in play but also involve conflict. There is an ultimate pay-off possible; to provide intimacy and finally its ultimate implication, sexual intercourse; or alternatively an excuse for its avoidance that can be blamed on the other partner. Either way there is a serious intention to win. All or nothing because of the others fault. In previous stages reassurance is sought and given, here the same reassurance may be sought and given but then turned back on the giver. These games are not only directly sexual but are also concerned with the postures that are taken up as life-styles, they are life games or roles that are being played out as well, and can become life long careers.

Life games are generally concerned with providing reasons for failure or why people think that they are like they are; "they drink too much" or "please won't someone help them;" or "somebody or something else made them like they are;" or those who asked to be kicked and get pleasure from it although they plead that they are asking not to be kicked; or those that blame everything on somebody else, their parents, bad schooling, bad housing, bad debts.

Marital and sexual games are played to fight off sexual impulses. They are a perversion of sexual instincts in which satisfaction is displaced from the sexual act to the crucial transactions which constitute the payoff. Their function is as a barrier to intimacy and the playing of another's games. The objective is to refuse the desires for intimacy of the other by making it look as if it were their fault while secretly being unwilling or frightened themselves.

Games are both necessary and desirable; not only do they structure

49

time satisfactorily, allowing much more variation and inventiveness than stereotyped ways of passing time in the area of social public games; but they maintain psychic stability for each person at a level that they find that they can go towards complete intimacy and for as long as they are prepared to go. People pick as friends those who play the same games or complimentary games that allow them to play the games that they want and need to play. These positions are taken up very early in life between the first and seventh year, long before the child is competant to judge or experienced enough to make such a commitment. Unless something or someone intervenes at a later date the rest of life is spent stabilizing these positions and fending off situations that threaten it.

From this point of view, child rearing is essentially a period of time when the child is taught what games to play, and how to paly them. The parents teach the procedures, rituals, manners, for his position in the culture. His knowledge and skill will determine what opportunities and success will be available to him all other things being equal. In this way the study of social dynamics cannot be separated from cultural investigation and individual motivation.

The position of a child or adult becomes fixed in the society by four different positions:

Historically games are passed on from generation to generation and have a strong tendency to inbreed along with the people in the family who play the same game.

Culturally the games are self-perpetuating, being passed on by the parents who belong to different social classes and pass down the games that their class favours.

Socially, games are between the boringness of preliminary games and rituals and the strict circumspection of close intimacy and therefore fill the waking hours with social intercourse that avoids boredom and exposure to excessive and difficult to handle intimacy.

Personally, games are played with friends associates and intimates who play the same games; social circles are tight and seem foreign to outsiders. To change games is to change groups.

Three noticeable different changes of posture are revealed in observations of spontaneous social activity. The changes appear in viewpoint, voice tone, and volume, vocabulary, physical gesture and the use of space. They are also accompanied by shifts in mood and feeling, ranging from excitment, passivity, gaity, and depression. These changes and differences give rise to what has been described as 'ego' states corresponding to those of a childlike state that reflects the mind and activities of a young child; those that reflect parental figures; and those of an adult capable of objective appraisal of reality.

CHILD

The CHILD is not childish but child-like; that is on the one hand the child who accepts everything the parent says and does as it is told, and on the other the rebel, intuitive enquiring, and spontaneously enjoying. The craving for novelty and new experience and the taking refuge in the familiar and ordered is the never ending battle of exploratory behaviour. It provides the insatiable curiosity and experimentation that leads to new experiences whether in childhood or adulthood and is the basis of creativity and the source of new conjectures. The need for exploration and re-intrenchment relects in the birth of revolutions in new ideas and investigation and stabilization which builds up awareness and experience. The child state is the source in each human being of charm, pleasure and creativity.

PARENT

The PARENT has two main functions; it allows an individual to act effectually as a real parent teaching a child the necessities for survival of the race and secondly makes many responses to the environment and other people' so automatic that a great deal of energy and time is saved. Routine matters whether driving a car, pot training, eating and so on. The parent looks after routine matters freeing the individual from making trivial decisions conciously and accepting that routine are done that way because they are done that way until an unusual situation is encountered when there is an immediate switch to a conscious decision making state called the adult state.

51

The parent is the teacher also giving directions about how things are done without expecting to be questioned or doubted.

ADULT

The ADULT is necessary for survival; it processes data and computes probabilities which are essential for dealing with the environment and it mediates between Parent and Child. It is the controller and conscious decision maker. A car is normallly driven after the skill has been learned, controlled by the automatic response of the Parent giving time for other states to operate but in a new situation or emergency the switch is immediately to adult level and totally involved. The Parent relieves the Adult from the necessity of making many trivial decisions.

All three aspects of the personality have high survival value and it is only when one comes to dominate the others consistantly that a problem arises. Otherwise each is entitled to equal respect and has a legitimate place in a full and productive life operating on all levels.

6

Learning and Spatial Ability

CULTURE, COMMUNICATION, LEARNING

Culture is treated here in its entirety as a form of communication; It is what is taught and what is learned.

All cultural behaviour has biological roots from which most, if not all, of it grew and is concerned with message systems about the way that the human organism has learned in the past to adapt to its environment and is relative in its content because of new learning being added all the time.

Culture can be experienced and be thought of as operating on three different levels and is communicated on these different levels both in social transaction and in the process of rearing successors. The alternation between three different types of awareness or conciousness imbues each experience that co-incides with the process of learning; from formal belief, accepting what is told; to informal adaption, modifying for different personal experience; to technical analysis, finding out exactly the purpose and reason for a belief; at whatever level or time scale.

LEARNING, INFORMAL, FORMAL, TECHNICAL

There is a connection between Informal learning and Childlike state of mind. Formal learning and Parentlike state of mind, Technical learning and Adult state of mind.

Informal learning activities are taught by the giving of a model for imitation. This way whole clusters of related activities can be learned at one time, without going into too much detail, and in many cases without the knowledge that something is being taught or learned at all, or that there are patterns and rules involved. This is the basis of many new theories of education concerning a seeming Parent to Child relationship but where the Adult control is carefully concealed.

Basic attitude: 'You will find out later but use your eyes to look around you now.'

Formal learning activities are taught by precept and admonition. The Adult moulds the Child according to patterns that he himself has never questioned. Formal lessons are learnt by making mistakes and having someone correct them by punishment or reward without giving reasons for the rules at any time. Parent to Child.

Basic attitude: 'Children don't do that and anyway you are not allowed to do it.'

Technical learning in its pure form is a way of learning process, transmitted in explicit terms from a teacher to a student orally or in writing and is in this sense the same Parent Child relationship in a more adult form. But more often the procedure is by logical analysis and proceeds in a coherent form so that each side understands each step, so becoming an Adult-Adult relationship.

INTELLIGENCE, VERBAL/SPATIAL ABILITIES

It has been found that while trying to derive tests for educational selection, that contrary to established educational beliefs, human mental abilities are spread over a wide spectrum and that this spectrum co-incided with the formal-informal-technical ways of learning each of which is more appropriate for each conceptual ability. It is clear that some find it easier to learn and to think one way about problems and their solution. An ability-learning hierarchy can be arranged as follows:

In a general factor of intelligence . . . .   g.

There are two major group factors . . . verbal/numerical   spatial/mechanical

There are further minor group factors. fluency . . . . verbal . . . numerical . . spatial . . .

mechanical . . technical.

The relative positions of problems and   priests   Accountants   Biologists   Mechanics

problems solvers approximately . . . . propagandists   mathematicians   Draftsmen

writers   Artists
Architects
Planners

On the basis of this finding and subsequent information following this theory the outcome seems to be as follows:

This hierarchy only holds in a free situation before formal education begins. The present educational system was evolved to train an elite of administrators, politicians, rulers, academics, and managers whose methods are based on a highly verbal literate, classics orientated, cultured being, with great demands on verbal fluency. For those with outstanding spatial ability there is only one way to achieve higher education and that is by becoming verbally more fluent however much it distorts the previous ability.

Current educational selection procedures give much greater weight to verbal abilities than spatial abilities which means that the school intake is biased in this direction. A very high proportion of those with outstanding abilities are being debarred from advanced cources where their ability is most needed.

Tests for verbal reasoning, numeracy/arithmetic, and English have positive value for linguistics subjects but a very negative one for much of science and nil for mathematics, natural sciences and mechanical-technical subjects.

Verbal ability or verbal reasoning tests called 'intelligence tests' were once believed to provide a measure of 'abstract intelligence'

The new electronic revolution needs people of great spatial ability

55

and spatial tests for mechanical ability or 'concrete intelligence.' It is now believed that spatial tests may be better measures of ability to think abstractly.

Spatial ability is increasingly important in advanced mathematics, abstract and analytical thinking and advanced problem solving. These are the abilities needed to work on the frontiers of knowledge. Information input for spatial types may be quicker through visual images than through print-verbal reading, lecturing and so on.

There is a clear connection between personality types and verbal and spatial abilities.

Verbal types tend towards verbal reasoning, verbal fluency, vocabulary; when discussing any configuration will pay a great deal of attention to detail over a small area. There is a tendency to narrow mindedness, mania depression, non-athleticism, emotional instability, adventurous extraversion, hysteria, placidity, femininity and dependency on locating themselves in space by reference to the outside environment; cannot visualise easily.

Spatial types tend toward spatial, drawing, mechanical, and practical; when discussing a configuration they tend to pay more attention to the whole at the expense of detail; they think big. They are more open-minded schizophrenic, wild and athletic, withdrawn and introverted, obsessionally anxious, paranoic, behaviourally alert, masculine and independant of locating themselves in space by reference to the outside environment; can visualise easily.

There is a parallel between verbal and spatial ability and the concious and dynamic unconcious mind; about one part above and eight parts below the level of consciousness. Abilities of the mind are essentially bi-polar. It is still a fundamental assumption that the verbal-conscious one eighth is what really matters in education and work and that the spatial-unconcious seven-eights can be safely ignored.

It is now thought that spatial abilities are in some ways more fundamental, more basic, and more dynamic than verbal ability and that the lack of verbal ability as a means of communication is becoming less of an obstacle with the introduction of models, television, diagrams.

The new electronic revolution needs people of great spatial ability with the ability to think abstractly and analytically together with a skill to visualise relations rather than separate and categorise them. Educational establishments are still geared to meet the needs of a managerial-administrative culture which places low value on spatial-technical abilities, that are thought to be appropriate to artisans only. A large amount of mental resources are being lost through dropping out of this system.

The rapid expansion of the information available in the last twenty years has come about through the development particularily of electronic technology and had enabled man to surround himself with information about the whole world in which he lives. The disassociated role of the literate man is becoming a much more participative and in depth role as awareness increases of other people and other problems. The greater the literate values are held in respect, the more difficult it is to see that things can and are being considered in a different way. These differences are most noticeable in the young because of their contact with the new electronic media at an early learning age.

It is in the area of education that the dichotomy between verbal and spatial values is most noticeable. Those in control of educational policy were educated and are the defenders of the existing culture based on literary values and past technologies. Instead of helping the young, who are more aware than in control of the new rising and revolutionary environments that are in the process of creation, education is being used instead as an instrument of aggression by trying to impose values that will not help to solve the new problems as they arise. The policy makers seem to be totally unaware of the consequences of their own discoveries and creations, apparently wishing to freeze the world in the image that was taught to them. The rigid parent educating the engineering child as has been mentioned before.

The generation gap is not so obvious a phenomenon as it may seem at first, where the gap really lies is between two cultures whose members might be at any age but whose outlook differs over goals. Education as it is now constituted is totally backward looking, an outdated system, seeking out an elite, whose data and values are too literate and fragmented and classified, being totally unsuited to the needs of a new generation whose information threatens to swamp them.

Todays children are growing up absurd and confused because they are suspended between two value systems; one based on the verbal-literacy needed to run a mechanical technology and the other on the values that seem to the child necessary to cope with the world that he sees around him.

The challenge of the new era is simply the creative process of growing up as opposed to the mere teaching and repetition of facts which are quite irrelevant to this process. There are too many things that are different in the electronic world of children for them to be expected to respond to the old education modes. Environments and cultures change with different sensory perceptions. The topics within the present education system are not in the same environment or culture and are therefore incomprehensible.

The desire for involvement is clearly growing; and the man who is involved is the spatial man not the verbal, detached man who remains uninvolved.

Different people fit into different ranges within a spectrum of abilities; the politician almost totally verbalises, reducing his problems to neat categories and delivering his solutions as if they had unquestionable authority; his thinking is in verbal-literate terms and lies at one end of the spectrum. Then come the numberate clerical scientist, the spatial draftsman, and the illiterate mechanic, who can put things together and take them apart with literacy, as can be seen anywhere in Africa wherever the literacy rate is low. Mechanical technology can be operated by people who almost totally illiterate as the development of the industrial revolution only too adequately illustrates. It took place almost entirely without help from the universities, who at the time were producing literary humanists and classics scholars not even numerate administrators, for managing colonies at best, while the engineering and designing was being done by self-taught artisans. There is a need for both, but the balance needs to be restored to make use of all the potential mental resources in higher education.

One of the most interesting reasons why the universities are failing to stop the rise of the spatial total involvement is that information reaching these people from outside the education system is very often more relevant to the present world than that from inside. For this reason the possessors of high intelligence in the spatial area, are getting the information they need, although they were flunked out of the system by verbal tests and examinations, and so are becoming competant in a way that is becoming much in demand.

This information does not only come from television but from books, magazines, radio, and by travelling around more, being sent abroad by the army, and by generally living a fuller life made possible by greater expectations and more money to spend.

Television does undoubtedly give more information but it also allows a more critical attitude to be taken and fostered to official cultural values. The opinions expressed on television can be compared with a greatly increased amount of learned and real life experience and therefore become involving. This involvement leads to paying greater attention to what is going on and to participate to put things right in the real world. The Vietnam war on television seems to have had this effect, for these reasons, not because of blood seen in full color or some tactile sensations from the screen lines.

It is a matter of the sort of information that the viewer gets from a verbal message accompanied by a visual image that depicts the same event. A message contains more information the less it is structured. A verbal message contains less information than a picture that involves many message systems. A verbal message in this sense is in effect censored by literate cultural values and its limits as an information carrier. On the radio or with verbal news for instance, a point of view is given that often conflicts with a picture. If a person is told that a good man killed a bad man, there is no way of verifying this—fact and judging anything about it. But on the other hand to see pictures that depict both sides in a war looking frightened and bewildered, not really understanding what they are doing or why they are there, covered in sweat and dirt, not quite like the movies of war heroes, and not knowing whether they are going to be killed; these messages come over through the multitude of sensory pick-ups in a very immediate way indeed. The watcher

59

is involved whether he likes it or not. If he does not, then he demands that the pictures should be banned; but if he does not he goes on a march to try to get it all stopped.

Those, whose education and life have only been concerned with literacy values from books or radio, only hear the vocal commentary, blanking out the visual commentary as an experience that they cannot fit into their expectation, they refuse to see the evidence before them and only get stimulation from the 'correct' verbal opinions. They cannot understand that others get different information and that it adds up to the opposite view to themselves. The communication gap is complete.

One of the causes for the shift over from verbal abilities to spatial ones is the possibility for self-education for a great number of people who would have otherwise never had a chance as a result of an electronic technology. But the origins of this technology were concerned in the developing of a technology in response to the needs of those who were trying to solve numerical and spatial problems, especially in warfare and space, and the spin off in terms of home entertainments was almost incidental, as the amount of development on television technology as compared with computers only too clearly shows.

It is now becoming clear that there are areas of human enquiry and problem solving that can only be dealt with as a totality with pattern recognition, exploration, probing and pattern making; and at the same time there is emerging a class of people to help in this task who for the most part are coming from outside the education system or in spite of it, but this is not for everyone and would be a disaster if it were made to be so. Mechanical technology is by no means over and there must be people to run it.

The next stage of knowledge advance is going to need spatial not verbal people and the swing is going in that direction. The era of mechanical invention is slowing down and it is possible that most major mechanical inventions that are needed have been made and that a period of consolidation has come when work will be concerned with increased efficiency and reliability which will come from work in the electronic-spatial areas. It is sure that an education system that will be useful in these circumstances must concern itself with the whole spectrum of human mental abilities and not just one end.

The values of the culture are being put under pressure to change in a very fundamental way. The highly informed new spatial child finds it impossible to adjust to the fragmented visual goals of the present education system after having all his senses involved in his life outside school. The requirement is for further involvement in depth in a meaningful way that relates to this experience, and not to some specialised and irrelevant linear detachment and uniform sequential pattern. It is found to be very difficult to cope with a vast bureaucratic structure of values based on classified courses, individual competition and inhibiting credits. At a time when advanced thinking is coming to the conclusion that competition with nature will destroy it and us with it and that co-operation with nature and each other is the only viable way of existing; the education system is becoming even more committed to the isolated-from-nature and an all-out-competition approach rather than the Gestalt approach that is needed.

Reactions have to be of now not some 'rational' verbalised distant goal of traditional education and this means exploiting the whole spatial sensorium. Traditional means of analytical methods just cannot absorb all the information that is now available. No one can know all there is to know anymore.

The stifling, impersonal dehumanized, multi-university, that grew up as the final grotesque answer to the problem of stuffing the maximum of ever growing body of 'facts' into brain-washed-recepticles and testing to see how much was remembered; must eventually give way to autonomous self governing study colleges devoted to in depth enquiry. Learning must respond to the demands by the young to experience everything as a complex inter-play of people and events given meaning by those with greater experience than themselves in exploring the development of forms and multi-levelled relationships.

Few adults seem to appreciate or comprehend youth's alienation from a bureaucratic mechanised consumer world and an educational system that is there solely to fit them into pre-classified pigeon holes.

7

Moral Rules

MORAL RULES

Moral laws are those that distinguish between good and bad behaviour and these rules are variable. Morality is specified by culture and it depends very much on the position in a given culture and the whereabouts of the culture as to what the rules may or may not forbid.

All moral rules are conservative; all new situations are compared with previous experience and dubbed abnormal, that is to say immoral.

The content of moral prohibitions vary widely, not only between one society and another, but even within the same society, as between one social group or social class; or one historical period or another.

It is a very deeply held belief that moral constraints are shared equally by all humanity but this is not true and is a delusion. Perhaps the only moral rule that seems to have a general concensus is the prohibition of sleeping with one's mother.

One of the rules of the present game that has always been accepted without question is that ethics and morality are tied to principles that can never be doubted, that have always been considered as fixed as the workings of the solar system. Science has never been able to show that there exist any such principles which are even theoretically acknowledged by all human societies; and neither has any priest or philosopher. Ethical values are nothing but functions of the societies in which they originate. The question of what is morally good or morally evil has no meaning except in reference to the value system of a given society—and there are no absolute criteria by which that value system can be judged. It can, of course, be judged on the basis of another system, but there is no possibility of deciding objectively which of the two systems is morally better. If they clash, the outcome must rest upon which one prevails: the possessor of a technology extensive enough to overcome the moral defense of the other, or vice versa.

THIS PRINCIPLE APPLIES NOT ONLY TO THE RELATIVE STRENGTHS OF DIFFERENT NATIONS BUT ALSO OF THE CONSTITUENT PARTS OF NATIONS. THE MOST OBVIOUS EXAMPLES CAN BE FOUND IN THE UNITED STATES, BUT THERE ARE MANY OTHERS.

Moral rules are very simple to apply because they demand yes-and-no answers to all problems with the consequence that any ethical problem that cannot be answered this way is treated as if it did not exist.

Morality allows the fitting of what is actually observed with some model that has been taught to be expected, that is, how things are meant to be. As long as these basic patterns are recognisable the toleration for quite wildly deviant behaviour from the norm can be accepted in others. But when the deviation goes too far, confusion sets in, and there is little understanding of what is going on. The usual way to restore confidence in an orderly world is to classify this sort of behaviour as wrong or immoral, to claim blindly that such things never happened in a previous age, and then to push it out from conscious thought altogether.

Other people's behaviour is evaluated according to an arbitrary code that has been taught. The code about social relations which changes in time and space constantly.

But the situation at present is making it more and more difficult to rely on many of the existing moral rules. Science and technology have allowed man to understand the balance of nature but also how to frustrate it. The Judaeo-christian ethic, by putting stress on the fostering of individual human life whatever the circumstances, has raised moral problems that no-one seems prepared to notice. Is it a doctor's duty to save every life: the abnormal, the senile, or those in chronic pain? When does a foetus become a human being? Is there ever a time when preserving a human life can become immoral rather than moral?

Modern medicine has given doctors almost unbelievable powers to preserve creatures that nature would have previously destroyed, powers to alter the life prospects of the child in the womb, the personality of the living and the life span of the senile. But if these powers continue to be used in an uninhibited way while similar efforts are being made to cut down the birth rates, then the possible outcome is a very decrepit conservative society in which all the political and economic advantages will lie with the very very old. For there is already built into the system that at present seems perfectly respectable morally, that not only the old shall rule but that the influential and the wealthy should have use of the wonder medical technologies if there is not enough to go around. For it is very noticeable that the direction of medical research and medical technology is not towards the prevention of illness and decay in the young but the preservation of the already decayed or already sick. Medical emphasis is on cure not prevention, and the unanswered and undiscussed question is who to preserve, the plumber or the prime-minister, who to try the newest techniques on, of these two; it is better to have given one old man a few more years life however important at the expense, literally, of preserving the lives of many just born children who are just as likely to be important in their life time?

Moral reticence supports the orthodox intellectual attitude of scientific detchment. It encourages the refusal to get mixed up in practical realities. Nuclear research can lead to nuclear warfare; medical research can lead to population explosion; chemical research can lead to pollution. Scientific research is a source of good and evil, to ignore the effects of this work and leave it to others to resolve the moral problems that it involves is to refuse to accept the dual responsibilities that it proposes.

In the hungry, bitter end, human interference with the balance of nature will be self-correcting but it is surely an old ethic if in the meantime Christian morality should lead to the avoidance of children so as to preserve the lives of the maimed, the senile, and the half-witted; or to commit the Christian societies to preserve their morality at the cost of killing or maiming almost everyone in them; to set out to preserve these values at a social cost that denies them and to preserve the lives of the inhabitants in such a way that will lead to their deaths if they ever had to do so.

The obsession with death rather than life, with preserving the dying old rather than the living young; to preserve the past rather than to create the future. The whole world at this point in time is dominated by this ethnocentric Christian ethic based on these premises, but there is already some indication that even within the general confines there might be some questioning and these questions are likely to be asked more and more often in the future.

Section II

The Nature of the Environment

8

The Conceptual Beach

Searching amongst the flotsam and jetsom on the conceptual beach of our time can be found the idea that hard radiation coming perhaps from the sun, the stars, perhaps from unknown forces in the universe, whose workings are little understood, strikes from time to time in a way that obliterates or changes certain symbols in our genetic code. If the changes are beneficial or not harmful then the new code will speak louder and louder in each succeeding generation. So we have a salamander where we had a fish; or a man where once there was a lemur.

In the same way, the effects of hard radiation from minds that glow like suns and shine like stars or cry in words that are only partly understood, introduce new ways of looking at our existence, its purpose and workings. In the short period that records have been kept several mutations in our world view have taken place. Once they were introduced, the world that followed, as it was experienced, was never quite the same again.

These changing organisms or changing views of the world cannot be given additional status nor put into some hierarchy of values. Evolution puts no increased value on one thing succeeding another, it only disapproves if the change is destructive to adaptation to the environment, whether it alters quickly or slowly. The useful or the useless mutations survive equally. The useless disappearing into the genetic or written records like and appendix or a religious belief.

The fragmented record of these mutations remain in our bodies and in our minds, one layer piled on another. The reason why things

69

are the way they are now is entirely fortuitous. A collection of past and present events on the way somewhere else with occasional violent changes of direction.

WE ARE NOW UNDERGOING ANOTHER SUCH VIO-LENT CHANGE OF DIRECTION, IF IT SEEMS DIFFICULT TO ADAPT IT IS BECAUSE WE ARE VICTIMS OF OUR PAST VALUES. . . . THE HABITS OF CHILDHOOD PERSIST.

If a disciple of each major mutation in the world-view and its values was asked to describe the way he felt, the way he thought and the way the universe looked to him, then it might be possible to summarise these attitudes in the following way.

All these attitudes are now present in the traditions that belong to the societies of Europe and North America and have infiltrated into the rest of the world, mixing up with their edges blurred, their influence bearing on every decision and every problem that arises when trying to fit the environment into experience.

THE WORLD-VIEW OF THE PRIMITIVE REALIST

"EVERYTHING IS ORDERED IN THE WAY I SENSE IT"

"Everything is as I immediately sense it, see it, touch it, taste it, smell it, hear it. Essentially I feel that everything is inhabited by spirits, I am inhabited by spirits. These spirits rule my destiny in this life and the next, so I have to come to terms with them even if I cannot know them and I do so by ritual, magic and incantation."

"The superior man, that I look, up to, is the successful magician or witch doctor who knows the spirits and can communicate with them."

In all countries, cultures and societies in the western world even today there are those who believe in the ruling powers of the spirits; whose help can be involded or whose wrath can be avoided through magic words, incantations and ritual performances.

"MY AUTHORITY FOR EVERYTHING I DO AND BE-LIEVE IN IS THE SPIRIT WHOM I CANNOT DEFINE."

70

THE WORLD-VIEW OF REASON AND LOGIC

"EVERYTHING IS ORDERED IN THE WAY I SAY IT IS."

Began 650 B.C. approx. onwards.

"There is a world, and me in a world and me observing myself in a world. I classify things, qualities and actions in the world and in me so that I can order everything. I use these classifications to guide all my behaviour."

"The superior man that I look up to is the one who is most 'objective'; whose thinking is as orderly so the world is ordely so the world is orderly. My brain is a mirror of the world; all orderly brains understand that each thought corresponds to a 'fact', each word to a thing, a person or an action, or a quality. My thoughts preceed logically from one fact to another. In my head is an exact minature of the world outside."

There are many who still think like this. These are the literate reasonable people who reach agreements by logical arguments and assume that all people agree with the basic premises. If they do not then they are call 'ill educated'. They rely on the metaphysicians to provide them with 'correct' principles so they must be detached from the corruption of normal experiences in the world. They are expected to construct perfect worlds that we must all aspire towards. These are the accumulators of 'facts' pinning labels on them, cataloguing them, basing their conduct on them and their appraisals on them so that there is a label for everything. All should be tidy and in its proper place. This is the basis of Western middle-class values and actions and so-called education. The language of the liberal, the politician, institutions, do-gooders or social reformers.

"MY AUTHORITY FOR EVERYTHING I DO IS REASON."

THE WORLD-VIEW OF SCIENCE

"EVERYTHING IS ORDERED LIKE A GREAT MACHINE WHICH I CAN TAKE TO PIECES AND DISCOVER HOW IT WORKS, THEN I WILL KNOW EVERYTHING."

Began 1500 A.D. onwards.

"I do not need to confer with vague spirits, or mistake my own voice for that of Naure. I can ask Nature definite questions and get definite answers. Any questions that are indefinitely formulated or do not give definite answers are not valid or interesting. I make formulations from these answers and project my conclusions into the unknown where I can discover further hidden facts of Nature. This technique has worked perfectly in my chosen field and can lead to an enormous technological revolution for those who try it. If you only apply these rules and procedures to all kinds of enquiry then we shall eventually know everything."

"The superior man that I look up to its the experimentor-mathematician who expresses relations in formulae that reveal properties of things, even if I do not always understand."

All problems are really very simple if they are approached scientifically can even result in a man on the moon. These believers choose to ignore the point that the only use they had for men in that operation was as the most efficient form of monitoring the technology. The result is the technological Utopia of the managers, the government advisors, the 'experts,' the speakers of charts, and figures, the 'efficiency' men who base their conduct and the appraisal of others' conduct on the basis that this is the only way of looking at the world. The cycle of science-technology-production-consumption is the only way to organise the world, there cannot be any other way because this is what science can do; any other way of looking at it is non-scientific.

"MY AUTHORITY FOR EVERYTHING THAT I DO IS THAT I CAN MAKE THE MACHINE WORK."

THE WORLD-VIEW OF RELATIVISM

"EVERYTHING IS ORDERED BY PROBABILITIES THAT I CREATE BY MY WAY OF LOOKING AT THEM."

Began 1900 A.D. approx. onwards

"I find that the further I ask questions, the less and less the world seems like a giant machine. I even find it difficult to ask the right questions and the answers often baffle me. I find that the answer are always in the frame of reference to a world which I myself have created through centuries of observation and thought. The structure of the world is through my own postulates; the world is what I made it. My ideas appear to be relative to my own spacetime relationship with the cosmos and with all the single events that I choose for study. What were once thought to be spirits of Nature or 'facts' of Nature, or 'laws' of Nature I find to be gross irregularities in a world as I see it through my own inadequate senses and instruments. All I can seem to find to be are statistical averages that provide rough indications of probabilities."

"The superior man that I look up to depends on getting the answers I want from the questions he asks. It all seems to depend on what questions he asks and how, or if, they relate to my problem."

In high number problems it was found that solutions could be found if averaged out, the statistical average adding up to a probability. They were found to have certain orderly and analysible properties, like telephone exchanges, or the workings of economics or planning. This is future projection on the basis that history repeats itself and all that has to be found out is what, in general, went on before for our future projections. It can allow no innovations because this would invalidate the figures so laboriously built up; what can happen must be based on what has happened or was thought to have happened. Somebody will always get the answers he wants to questions that determine them; it makes it much easier to prove a point if someone has already determined what it is. Any frame

73

of reference gives the answers wanted so it can appeal to the pressure of any statistical group that will back any opinion; whether the tax payer; a particular majority; the homeowner; the motorist; the planner and so on. This is the world of the self-fulfilling prophesy.

"MY AUTHORITY FOR EVERYTHING I DO IS MY SPECIALLY CHOSEN STATISTICAL AVERAGE."

THE WORLD-VIEW OF WHOLENESS

"SUDDENLY I SEE THAT EVERYTHING HAS AN ORDER WHICH NO FORMULATION CAN ENCOMPASS. ALL PREVIOUS IDEAS EXPLAIN LITTLE BITS BUT CAN NEVER ANSWER THE CENTRAL QUESTION. I CONCEIVE OF THE WORLD AS MY OWN TOTAL EXPERIENCE WITH IT. I PLAY WITH MY OWN SYMBOLIC CONSTRUCTS IN A SPIRIT OF EASY DETACHMENT"

Began 1966 A.D. approx. onwards.

Heisenberg demonstrated that I cannot separate what I observe from my own act of observing. I find that my total organism with all its personal history, as well as my brain with all its history is engaged all the time. I have discovered that all my cleverest ideas have their origin and meaning from sensory contact, fusion and oneness with what is going on around me; beyond the symbols that I have been taught; and without distinction between self and non-self. I find that everything is inter-related into an organic whole. The behaviour of things as a whole is the more interesting than separating them into parts. Their existence is dependent and most useful in their inter-relatedness and the certainty that they cannot exist interdependently. The meaning of life is in everything from cosmic energy to fellow human beings equally and that verbal distinctions between "art" or "science" or "practicality" disappear becoming an overall oneness of . . . experience. The observer and the observed are one. I communicate with my world as no other

74

can; I exchange information with it enhancing it and myself; the greater the exchange the greater the involvement. What is important about the statement 'one and one make two' is the 'and' not descriptions of 'one' and 'one.' This is a vital distinction."

"There is no superior man to look up to; each is an individual to the maximum of his experience in the world."

There is no authority for beliefs and actions, there must be a learning from mistakes as any other organism must, and change accordingly. The world is different for everybody and it cannot be seen in any other way. Each must do what he thinks best as long as it does not interfere with the rights of others to do likewise. But he cannot know everything and must learn to collaborate with others to achieve things. Everyone who is a product of the same culture, the same child-prejudice can only operate within its confines. Everyone cannot forget that which they have already learned, they can only select that which they want or need; they can only try to be themselves . . . freedom is to be oneself.

"I CAN AND MUST DO MY OWN THING."

9

What We Know
and What We Don't Know

PHILOSOPHY

Present Western culture has a heritage of idealistic philosophy, of principles and utopian visions. It stems from the dichotomy of a world divided into external things that are thought to have no value, and the inner world of human thought and reason to which alone values are attributed. There is always a comparison between the "idealist" and the "realist" which implies moral values where what is explained by natural laws is automatically thought to be devoid of value also.

Furthermore, this philosophical tradition is authoritarian in structure even in its most liberal form and it contains two central dogmas:

—The assumption that 'criticism' necessarily means finding justification.

—The assumption that the quality and degree of 'rationality' proceeds from a justifying premise to a justified conclusion. That is to say, proceeding from an assumed premise to a logically justified conclusion is a valid way of increasing knowledge.

Authoritarian in structure means that philosophical questions are always put in such a way that they beg authoritarian answers like:

77

"How do you know?"

"How do you justify your beliefs?"

"With what do you guarantee your assertions"?

Only some authoritarian source can guarantee the correctness of an assertion; whether the appeal is to God, to experience, to knowledge and so on.

Sir Karl Popper has proposed a way out of this dilemma. He suggests that rather than ask for authority, which we known can never be ultimately challenged and could always be wrong, we should try to counteract unsubstantiated beliefs in our views. This has been called 'comprehensive critical rationalism.'

—Comprehensive because nothing must be left out.

—Critical meaning questioning all sources and ideas.

—Rational because it must be done systematically.

So questions like: "How do you know?" can only be answered: "I do not know; I have no guarantees: I can never know for certain."

Intellectual life should be arranged together with all institutions, in such a way that it exposes all beliefs, conjectures, policies, ideas and traditional practices, whether justified or not, to a maximum of criticism to counteract and eliminate as much intellectual wrong headedness as possible; in this way standards should always be criticised and corrected if found wanting in a continuous evolving form rather than once and for all standards based on a consensus of one particular time.

CRITICISM

Rational proof of a rational standards must be rejected because any authority appealed to cannot be shown to be more justifiable than any other. Rather than an endless search for infallible authority we should learn from our mistakes; this is what should be meant by criticism.

This is the non-justificational view of criticism; we should not try to cut out those things that do not fit within any theory; a fallacy called "induction"; or to try to find a theory which fits everything; the fallacy of "liberalism" and reduction to inaction; nor is it to fit everything into a preconceived mould that gives a particular meaning and purpose to everything; the fallacy of "totalitarianism" and "utopianism."

All necessitate that a premise is required to be acknowledged by another premise that cannot be necessarily accepted. To criticise a position in these cases it must be shown that either it cannot be derived from, or else conflicts with, rational authority, which in itself is not open to criticism.

The situation is really a very simple one:

It has always been believed, certainly since the Renaissance, that however much the standards fall short of reality, the truth will always manifest itself to any enlightened man however veiled it may be by the restrictions of the society in which they may at any time live. There has always been this optimistic view that man has the power to discern truth and acquire knowledge about it. It is not easy to see but once it is in front of him all men have the power to see it and to distinguish falsehood and know the truth.

Modern science and technology are based on this idea; the idea that everything could be comprehended if taken to pieces. There was no need for authority as each man could find out for himself. The other view that was taken was the view of human depravity that one man who had seen the truth must try to force it on the others who have not been so fortunate.

SOURCES OF KNOWLEDGE

But sources of knowledge are everything from anywhere, none has authority, and their origin has nothing to do with their validity. It is a fundamental mistake to confuse questions of origin and questions of validity. The only test is what has been asserted, not where it came from. "How do you know"? does not question sources; if it did, it would be meaningless. So the question is accepted and

79

authoritarian justification found for it. Yes and no answers are asked to questions whose validity has not been challenged nor their legitimacy. Questions like "Who should govern?" or "Whose policy should be adopted?" should be replaced by questions of how institutions can be organised best, assuming that they are needed; so that the bad, the incompetent and the ignorant can do the least harm? What is the best way to avoid or detect and eliminate unsubstantiated beliefs? What kind of mechanisms can be adopted to serve as a feedback to check mistakes?

Questions like "How do you know"? must be answered; "I do not know, my assertion is merely a guess, and instead of sitting in an arm-chair disputing our respective views all day, let us go out into the real world and see who could be right or see in what ways we have to modify our views until they fit better. To think something in your head does not make it exist, or even make it possible to exist. We can only learn from our mistakes."

CONJECTURE AND REFUTATION

It is Sir Karl Popper's view that philosophical thought, as any other problem solving for any other subject under discussion must consist of a rapidly alternating and interacting imaginative episodes of thought, a conjecture, a guess, guided by the knowledge of the time; and a critical episode of thought. An opinion is formed, a view taken, a hypothesis framed, an informed guess made, about the truth or the solution to a problem. A possible world is invented, a story told; all which lie outside logic. These imaginative conjectures are subjected to ruthless, rigorous criticism to see if they correspond as a first approximation to the real world. If they don't they must be absolutely rejected.

It is the critical process of testing that is important; what is being tested is the logical outcome of the hypothesis. If such a conjecture fails to be refuted then for the time being, and only until such a refutation is successful, can it be assumed that additional knowledge has been gained.

PROBLEM SOLVING

The process of conjecture and refutation provides the only definition of problem solving that will withstand its own rigorous form of criticism and provides a starting point for any investigation into problem solving methods.

It suggests that by bringing out mistakes it makes it possible to understand the difficulties of a problem that the solution is being looked for. It is in this way that a problem can be known better and therefore more mature solutions can be offered. The very refutation, and this is more likely than not, of a theory or a hypothesis leads a little nearer the truth or the re-solution.

KNOWLEDGE

Knowledge; which is information-learning stored in a symbolic form either in the head or some other external way, is the result of tentative solutions to problems brought about in adapting to the environment. These solutions are controlled by criticism from past experience; past mistakes. What gives rise to new problems is new information-experience in the real world, the environment.

All our knowledge grows only through correcting our mistakes.

CRITICISM IS NOT FINDING JUSTIFICATION BUT IS THE PROCESS OF FINDING TRUTH.

The rightness of an answer to a problem by cross examination from knowing and experience. The greater these are, the greater the value of the criticism. It concerns questions asked of the premise as much as the logical deductions from the premise. Its purpose is to counteract beliefs only substantiated by justification.

CREATIVITY

The imaginative process is not so intelligble as the critical one. The formation of new ideas, conjectures, is not well understood. But there is nothing to show that the process is any way different for solving problems in philosophy, science, art or any other area of problem solving. It is very similar to other intuitive and inspirational processes, including the basic one of initial learning in an infant where answers are sought to the question: "What if . . . ?" If it is assumed that this is the definition of the word 'creativity' then it can be assumed that all people are creative and the possession of creativity is essential to remain alive by constantly solving real world problems. What is important is the amount of criticism the conjecture is submitted to, both before and after it is put forward in a conscious form. It can be assumed that any hypothesis/solution has already been subjected to a fair degree of criticism before; the amount is dependent on the knowledge of the proposer and this process of prior censorship is called 'judgment.' What is worthy of note is not creativity but the critical faculty both before and after a proposed solution.

PROBLEM SOLVING

THEREFORE, THE PROCEDURE FOR PROBLEM SOLVING CAN BE SEEN AS THE CREATION OF POSSIBLE ANSWERS TO A PROBLEM AND THEN SUBJECTING IT TO CRITICAL REFUTATION. IF SUCCESSFUL, THE ANSWER IS SUFFICIENT ONLY UNTIL A REFUTATION IS FOUND WHICH WILL APPEAR IN THE FORM OF ANOTHER PROBLEM, STARTING THE CYCLE OVER AGAIN. NOT ONCE AND FOR ALL ANSWERS TO ONCE AND FOR ALL QUESTIONS.

KNOWLEDGE, THEN, IS THE RESULT OF RE-SOLVING PROBLEMS: PROBLEMS THAT ARISE FROM OUR COPING WITH OUR ENVIRONMENT.

It is learnt information stored symbolically: in the brain, in books, in machines, in buildings, in all culture.

As time passes there are many more of us and we know much more. The central problems of one era become the conventional solutions of the next. They become incorporated into the General Knowledge becoming further disseminated over a wider and wider field as time passes. At any one time there are those who are trying to solve problems to which no one has yet found an answer. As the human body develops from simple cells and gradually builds up in complexity, so does knowledge for each individual. Each person has to solve the problems for himself. The storage of information allows him to seek the solutions already made to his problems so cutting down the time needed to get into the problems that are yet without solutions. It is all part of the learning process.

But the process must be gone through, there is no way to short circuit this fact. For some, there is no need or desire to reach this point; they are content with the amount they know and find others who are in the same position. In this way, groups tend to congregate in the same place, or small towns, or a bridge club. Each area represents a 'cultural' group. Each represents different stages of knowledge at different stages of time and development. At one end the more esoteric 'impractical' concepts; at the other the more practical 'down-to-earth' ideas of the 'realist'.

The total sum of these ideas and attitudes is called culture. It could be described as an ever expanding triangle of knowledge. The particular, individual, unaccepted small voice at the top; and the general, mechanical, accepted mass voice at the bottom. Each part of the cultural whole can be looked at this way. At the top of the triangle there are those who are concerned with the problems of more complex flexible and less predictable patterns of activity. At one end the emphasis is on the 'content' of the activity, at the other the 'organization.'

10

Science as a Branch of Knowledge

Scientific discoveries form the groundwork for the technology of this time; but it is also instrumental in changing all the concepts of what man thinks about himself, where he came from and the world around and about him. If the future is to be understood or the present thought about, then the more salient discoveries and methods of science must be understood and used.

To understand the nature of the world is one step to understanding man's place in it. Science will not reveal what life means only how it seems to work. It is a general collection of knowledge that is generally agreed on, and which is ever enlarging, tentatively held and in a form that can be independently checked and reconstructed by anyone familiar with its methods.

The orthodox view of science is that the world is an immense machine that can be discovered how it works by taking it apart. The superior scientist was the experimenter-mathematician who could reveal the properties of men and things and their actions in easily followable sequences.

THE FALLACY OF INDUCTION

It is a common and false assumption that the scientific method consists of writing down all the observations that can be made about an aspect of the world, quantify them, and assume that a 'creative leap' would explain their interaction that could be embodied into a scientific law. This is the process of so called 'induction' which starts from the simple declarations of fact which are supposed to put on record the evidence of observation, evidence of the senses, and by logical exercises and operations, can build up these singular statements into general statements, and these general statements into laws of Nature. Following from this, there is also the opposite procedure where it is thought that if a complex and far reaching observation is made and then broken down into small enough small pieces, then somehow its relation to the laws of nature can be immediately seen.

This method is fallacious for a number of reasons but has become the model for most problem solving techniques in architecture, management, and other quasi-scientific areas of enquiry.

Inductive theories fitted well with the classic belief in the powers of reason; it found reinforcement by the 'great' single discoveries with their apparently far reaching repercussions, in spite of the fact that such ideas were later, and inevitably, disproved; nor did it give any due weight to the fact that most scientific research leads nowhere; or to false conclusions; or to thought processes that lead more often than not to error than truth.

THE THEORY OF KNOWLEDGE

But the main reason for doubt lies in the domain of the theory of knowledge. It is said that inductive thought starts from simple declarations of matters of fact which are records of plain observations. They are reports of the evidence of the senses, either directly or through instruments, and these are believed to have a very high degree of trustworthiness. But the question does arise as to how much such observation can be made without prejudice or bias, or preconceptions of any kind. The difference between fact and fiction is by no means as distinct as it would like to be believed. Facts cannot be observed as facts except in virtue of the conceptions that the observer himself unconsciously supplies. Life experiences as well as inherited ones are included. More than anything the basic notion of objectivity and subjectivity are thrown into doubt as a possibility of discovering the truth. Truth is not manifest and experience is valuable.

REASONING

There is a feeling that the process of reasoning is logical in the strict sense like mathematics. That the operation of thinking is something that is conducted with words, numbers, and symbols, and that there is a close interplay with external reality, and that language is as it were a system on its own, not just an instrument. But when talking about an idea or a concept, the idea or concept is conditioned by the words, and the meaning that is given them by the user; as the description continues the words, are apt to take on new meaning. The same words are being used but they have acquired a new content. Language has its rules of grammar but it talks about concepts and what is interesting is the concept.

The same can happen when trying to understand problems, the very act of trying to describe them with words alters the problems; it would seem that spoken language is inadequate to state, let alone solve, problems, as they can never be pinned down this way. They may exist at one time but change as time passes. The frame of truth, the answers to problems, is always altering as the understanding of them increases.

From this stems an operational point of view that always, wherever you go, there is more and more and more and only the edge of an unlimited universe is being explored, the sum of all knowledge cannot exist and therefore will never be reached, and that truth itself, as it were, changes as you look at it. This is a very relative way of looking at things. The concepts mean that no authoritarian statements of principle, ideal or utopian vision are acceptable because they do not allow for a continuous rearrangement and evolving. The problem depends on what you are experiencing.

THE SCIENTIFIC METHOD

There is another quite different and more viable conception of the method of science which coincides with Sir Karl Popper's view of knowledge in general, according to which the forward movement of science cannot be mechanised. His view is that scientific knowledge cannot differ in any respect from any other sort of knowledge and its acquisition must be by the same method of conjecture and refutation. In this way science becomes a way of thinking, not a collection of facts or certain or well established statements, nor does it ever advance towards a state of finality; nor can it claim to have attained final truth. It is a useful instrument for making the maximum use of the senses, not a mystique but based on the idea that mistakes can be learnt from and problems resolved, but every solution only gives rise to more new problems.

The method of conjecture and refutation is applicable to all problem solving and it was extremely unfortunate that the method of induction had apparently sufficient resemblance to be defined as the only method of problem solving. There was no suspicion that there was an incorrect definition because the method itself determined which direction enquiry should go by being successful in very particular areas. The study of things, the environment, led to a vast and highly developed technology, that it laid waste and destroyed large amounts of the world's surface did not seem to matter at the time and so consequently it was assumed that a similar study of men would lead to equally successful conclusions. It is here that the flaws in the methodology began to show and which were exposed even in the study of things.

The inheritance of the world of reason still prevails in the world of science. The categories of thought devised in the 5th century B.C. were still thought to operate. Rational analysis allied to the correct slotting of experience into intellectual areas like law, philosophy, ethics, literature and other notions of civilized living. The final culmination being the two-way split between 'science,' the world of things; and 'art' or 'humanities,' the world of people: the world of objectivity and subjectivity. The use of language as a linear form of communication and the resulting methods of education as if there were only sequences of events and tight compartments led to the dead-end and discussions of 'two worlds' apparently at war with each other.

BEHAVIOURISM

To make the methods that apparently worked in the environment so well for the study of man it proved necessary to look at him as if he were a machine as well that could be taken to pieces. So an orthodoxy appeared in the life sciences that became the background to the study of man's behaviour based on the idea that that which could not be isolated and classified did not exist, anymore than that which could not be isolated as part of a linear deterministic cause and even progression like the operations of a machine and Netonian physics.

It has become part of a basic orthodoxy that:

Biological evolution is a result of totally random mutations preserved by natural selection: if enough monkeys keep going on typewriters long enough they will write Shakespeare.

That mental evolution is the result of random tries preserved by reinforcement rewards: the carrot and the stick attitude to learning.

That all organisms, including man, are essentially passive automata controlled by the environment, whose sole purpose in life is the reduction of tensions by adaptive response.

That the only scientific method worth the name is quantitative measurement, and consequently that complex phenomena must be reduced to its simplest elements accessible to such treatment without worrying whether the specific phenomenon, for instance, man, may be lost in the process.

This crude slot machine description, the culmination of the Pavlov experiments with dogs, and Skinner experiments with pigeons, now permeates present general attitudes to philosophy, the social sciences, education and psychiatry. The use of the word 'conditioning' with its rigid deterministic connotations has become a key formula for explaining human behaviour and why and what they are.

MYTH OF QUANTIFIABILITY

Central to the theory of 'behaviourism' is the myth of quantifiability that suggests that all behaviour that cannot be taken to pieces and described as a stimulus-response sequence cannot be allowed even if it means that the idea of mind itself has to be dismissed. This is the simplistic absurdity of eighteenth century mechanism extended to attempts to construct a theory of human experience and mental activity on the model of Newtonian physics, which had apparently worked so well, and which now can be seen as theoretical lunacy. It is very unfortunate that such mistakes are now appearing in systematic form in social planning diciplines like architecture and planning.

In order to make the totality of experience intelligible it seems that it is necessary to redefine, reclassify and redescribe it to a point that renders the subject of the speculation utterly beyond recognition as reality. It is assumed that 'pure behaviour' can be described as physical or verbal actions completely separated from any attitude or intention or thought; what a person did as distinct from what they thought or said they did. To wave the hands above the head cannot be separated from the intention to hail a taxi or to fend off an attack or participate in a magic ritual. It is the intention of what people do that is interesting not the most obvious gesture that accompanies it, for this may be only the most obvious observation to call out or to wave a leg may also accompany the waving of an arm. To talk as if physical action alone can be regarded as isolatable aspects of human experience, let alone all the other factors, is not to talk about human experience in any meaningful way.

By making the content of human lives unrecognisable and redistributing them according to arbitrary categories it would then seem that these preconceived notions are then proved by an outrageous procedure which is to provide statistical evidence to prove the original supposition. Questions in statistical surveys can be phrased in such a way as to build in any researchers preconceptions into the only possible answers. They are highly useful to the theoretical frameworks which function as self-fulling prophecies in which the conclusions are determined by the nature of the research techniques. Such an argument is a particular favourite of politicians as well as social planners who support their views with appeals of majority backing without necessarily saying a majority of whom or whether an even larger majority could not be found for a proposition that they had not mentioned or discussed. Such conclusions can only be wrong or trivial; they certainly cannot be operational.

At the time that the dead end of behaviourism was being realised the overall effect of new concepts in the physical sciences were beginning to be felt. It was becoming almost impossible to maintain the category system of enquiry in the face of overlapping use of concepts and instruments; astronomy only started developing again rapidly with the use of electronics and nuclear physics; biology took in chemistry to make new break-throughs. It became more and more obvious that a separation into categories even on the scale of "art" and "science" was becoming nonsense; musicians began to use electronics and nuclear scientists pattern recognition and games as their basic tools. It was realised that everything was intimately interconnected and to look at one part of the system was to look at a part of the whole without boundaries; that nonco-operation lead to destruction of the environment and co-operation led to forward movement; that there was no possibility of something for nothing. By using words, names, categories in order to make sense of experience only ensured that they were kept apart.

It has become obvious that questions about how to partake in what is going on around, or whether we could be uninvolved observers is not only irrelevant but is meaningless. We are part of nature and to destroy or tamper with nature is to do likewise with ourselves and that this is what Science is describing and exploring; a totality of which we are all a part. The total explanation of an animal is not just what it is made of but also the way it connects up with its environment; the way it modifies it and the way it is modified by it. So far a lot is known about the environment and about the animal but little about the connection which we call behaviour. Behaviour is not determined by genetic endowment, which merely sets limits on what can or cannot be done: the question is where does man fit in and what is his relation to all the rest.

THE ANSWERS LIE IN THE BASIC CONCEPTS OF THE NEWLY EVOLVING WORLD-VIEW OF WHOLENESS.

11

The Changing World-View

Suddenly I see that my world has a structure that no formulation can encompass. All previous ideas explain little bits but never answer the central question. I conceive of the world as my own total experience with it, and I play with my own symbolic constructs in a spirit of easy detachment.

Having discovered that, I cannot separate what I observed from the act of my own observation. I begin to study my own way of observing. When I do this, I find that my observation does not consist solely of what goes on in my brain but that my total organism, with all its history, is also engaged.

Out of the following knowledge comes an awareness of my interrelatedness with everything; and the old verbal distinctions between art, science, religion and all the others disappear becoming an overall oneness of experience.

It is the comparison of relations that is in the end more thought provoking than the comparison of things, explanations more interesting than probabilities. Science is now more concerned with how things work rather than what things are; it is more concerned with relations than with objects.

It has up to now been found easier to think of relations as if they really existed and were not a by-product of a way of thinking derived from the peculiarities of the way language is formed and transmitted. Even in particle physics, although much of the experimental evidence and concepts are concerned with relations and all entities hypothetical; the idea seems unavoidable that the existence of relations must imply the existence of things that are related. A neutrino is by definition non-existent having no charge and no mass, and is most conspicuous by not being there but it is still given a name as if it were.

Speaking together is only possible because in spite of each private language which has been personally developed to understand personal experiences, with no proof that the same thing is meant by others, occasionally it is possible to make comparisons between relational structions as distinct from material things.

From this relational point of view any two machines or other entities that work the same way or do the same kind of job are the same kinds of machines, even if they are made of quite different substances and operate in quite different environments. In this way one machine can be compared with another just as usefully as the shape of a whale and an aeroplane. The shapes are similar for maximum efficiency when passing through the machines for providing forward motion are quite different in their energy source and the way they are built.

All things are inter-connected and the form of man's perceptions of these things is a result of the way he is made. As things are looked at in greater detail they seem to slip away. The only reality with which any real connection can be made is the pattern of relations. Things as objects are separate and so do not therefore exist; but relations connect everything up.

The world outside depends on how we react to it, the rules can be changed at any time, the point of reference altered, and nothing will ever be the same again. If we had different or more refined senses our environment would not only seem different it would be different; and now our senses have been altered the world is different. It is relations, that is pattern, which constitute existence, and what is recognisable about the external world is sets of relations; and not sets of real objects.

The alternative to chopping up human experience into arbitrary and counter intuitive ways as the behaviourist does, is to treat the human condition in terms of how it is actually experienced. It seems that it would be much more sensible to treat the condition psychologically, socially and environmentally in ways that result from human experience with no sub-divisions, categories artificially created to fit some pre-conception. Human experience is not composed of discreet actions, but is involved in them as it is involved in thoughts, attitudes, emotions and so on; outside this context, physical gestures are unmeasurable and meaningless. Each is related to the other and the total pattern only, is meaningful.

Experience is a Gestalt process, not a group of individual ingredients that can be separated out for the purposes of social or environmental planning. Such planning can only be in terms of how they are actually experienced and how they cause people to experience themselves and others; not the physical products. In this way, people come to the fore and cease to be a collection of actions and mountains of data that must be de-coded for the purpose of making something that physically corresponds.

To consider every human being as a point of view, a center of experience, is the classic existentialist point of view and is a great deal more workable and morally superior to the rats-in-mazes alternatives of the behaviourists. The ability to define personal experience and a person in terms of it, is the definition of sanity. Any environment that succeeds or tends to alienate or oppress an individual, as an individual subject of experience and that forces him into compartments that are depersonalised, demanding rigid conformity, constricting imagination and sensibility is directly responsible for pathological aggression and anti-social behaviour.

The human imagination needs an environment which permits a wide latitude as well as privacy and stimulation of the senses. Without these, the human being is incapable of experiencing himself as a viable individual. The individual must be able to see his perception of the world, and of himself, as valid and not have it perpetually negated, underminded and distorted by his environment which has been made to correspond to other people's perceptions or convenience of organisation. The reflection from the environment must be of a self that is not aberarant or grotesque to the beholder.

Section III

The Nature of Information

12

Entropy, Information, Feedback

ENTROPY

The degree of organisation of any system can be assessed in terms of the measure of probability call ENTROPY. The notion of entropy, originally developed in thermo-dynamics is a formal expression for the tendency for a 'closed system' to deteriorate or run down; going from a highly organized, differentiated and less probable state, to a more probable, undifferentiated and chaotic state. This process, which is symbolised by an increase in the entropy of the closed system is easily visualised if a physical system is thought of as composed of discreet regions that were originally at different temperature; as time progresses, the system will tend toward an equilibrium state, that is nothing new ever really happens. Therefore, it could be said that as temperature is only a manifestation of energy all the energy must have become dissipated and lost.

Both living organisms including human beings and machines constitute local pockets or islands of decreasing, time-limited entropy. There are indications of greater orderliness in a framework in which general entropy is assumed to be on the increase. This local reversal is caused by taking in energy from outside and so they can be described as 'open systems'. This energy source is most commonly called INFORMATION, which can be seen as similar and as useful a scientific concept as matter, energy and electrical charge. But the scientist's or the designer's quest for order in the universe, which is what we mean by problem solving, is always hampered by a phenomenen which can be interpreted as either a force contrary to order or the absence of order.

THEREFORE, TO EXCHANGE INFORMATION IS TO ORGANISE, IS TO REVERSE ENTROPY.

INFORMATION

INFORMATION is a basic concept which has no precise scientific or technical definition and can be many different things to different people. It is a 'commodity' that circulates in a communications system, no matter what its physical form. It is no more possible to define than energy, which is closely akin to the idea of electricity or matter, none of which have even been successfully defined.

There is usually a source of information such as a speaker or a writer who formulates a message which can convey an idea in his head. The message is phonetically organised in an oral or a written form, and through some medium like a telephone, a letter, a small boy, transmitted to the receiver. Here, the message is decoded from the electrical signal or piece of paper into a form which makes the idea clear in the head of the recipient. Providing the speaker and the listener have the same code, that is they have the same spoken language and belong to the same cultural community and providing the message has not been garbled by the medium by which it was sent then the reaction of the recipient in his return message will indicate whether the idea the information, was correctly received and understood.

As the coding-decoding device, the brain, is subject to mechanical limitations; the more effort the sender puts into organising the words of his message the less room there is for the idea he wanted to express and so the smaller amount of information that he was able to transmit. If the receiver already has some idea of what the other is trying to say to him then the message can be shorter than in a situation

100

where the idea is totally new. In this case a great deal more information is needed and so a great deal more time and organisation or words to explain the idea has to take place. The objective is to get an idea from the sender's head to the recipient's. This is in effect controlling the other person, making him understand and accept the idea. Once the awareness of the persuasiveness of this concept of information is realised it can be seen that the more information there is the greater the power for controlling the actions of any recipient.

Our adjustment to the world around us depends upon the information that our senses provide. It is obvious that information not only come from other human beings but the environment itself and that we feed information to the environment that like us is moving towards entropy. Our cultures depend upon the relevant use of the vast accumulated stores of information and the use to which we put them to reverse the effects of entropy. To know how successful we are in this task we rely on observing the effects of our attempts to control the environment or other people by phenomenon called FEED BACK.

FEED BACK

FEED BACK IS IN EFFECT NOTING OUR MISTAKES AND CORRECTING THEM.

Man's advantage over other species lies in his ability to adapt to radical changes in his environment through the use of physiological equipment that enables him to continue learning throughout his life span. Learning is the modification of behaviour on the basis of past experience by this same process involving the feedback of information. This concept consists of modifying the behaviour of any system by acting on the results of actual and not just expected past performances.

It can refer to the success or failure of a simple action or it may occur to more complicated levels when information of a whole policy of conduct or pattern of behaviour is fed back enabling the organism to change its overall strategic planning, or life style. The advantage of the concept is that it does not require the quantification of every single item and event in this cycle but only the measurement of the difference between what information is put in (in order to control); that which is expected; and the actual results. This makes it possible to take into account those things that cannot now be measured and therefore tend to be left out of the more conventional organisation theories and tools.

There is no need to depend on the maximum amount of information before anything can be done as this does not always have to be taken into account; it depends on the degree of sophistication and amount of response needed. The more the information involved the greater the sophistication. The usefulness of this concept can be seen when applied in areas where little quantification of parameters is possible, in the social sciences, psychology, etc. As long as the parameters or variables are capable of being formulated in some recognisable form with respect to the information that is required to be fed back then there is no limit to the functioning by learning from the consequences of correcting previous behaviour.

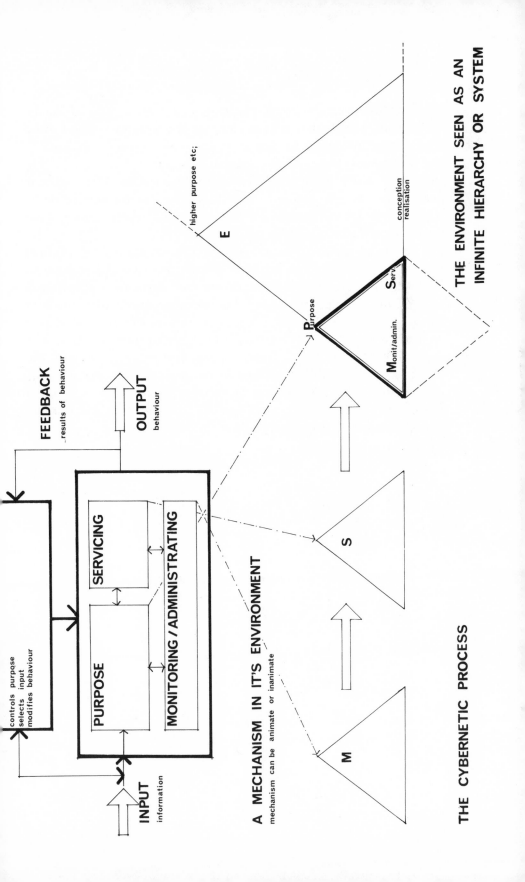

FEEDBACK
_results of behaviour

OUTPUT
behaviour

controls purpose
selects input
modifies behaviour

SERVICING

PURPOSE

MONITORING / ADMINISTRATING

INPUT
information

A MECHANISM IN IT'S ENVIRONMENT
mechanism can be animate or inanimate

M

S

THE CYBERNETIC PROCESS

higher purpose etc;

E

Purpose

Monit/admin.

Serv.

conception
realisation

THE ENVIRONMENT SEEN AS AN
INFINITE HIERARCHY OR SYSTEM

13

Control

Information is a name for the content of what is exchanged with the outer world as we adjust to it and make our adjustments felt upon it. The process of our adjusting to the contingencies of the outer environment and of living effectively within that environment. The needs and complexities of present day life make greater and greater demands on this process of information, and to be able to act effectively is to live with adequate information.

This communication and control belong to the basis of man's inner life even as they belong to his life in society. To be without adequate information is to be without adequate communication or control.

To have adequate communication or control it is necessary to know what messages carrying information mean.

PROBABILITY TENDS TO INCREASE AS THE WORLD GROWS OLDER.

A measure is called 'entropy' and its characteristic is to increase with time; as it increases the universe and all the closed systems in the universe, tend naturally to deteriorate and lose their distinctiveness. To move from the least to the most probable state; from a state of organisation and differentiation in which distinctness and form exist; to a state of chaos and sameness.

ORDER IS THE LEAST PROBABLE: CHAOS THE MOST PROBABLE.

But while the universe as a whole is tending to run down, there are local enclaves, whose direction seems to be in the opposite direction to that of the universe at large, and where there is a limited and short term tendency for organisations to increase. In these areas reside those organisations which are called life systems and their related social systems. What causes the reversal of entropy in these cases is the introduction of INFORMATION: making these systems susceptible to outside influences.

INFORMATION DECREASES ENTROPY AND BUILDS UP ORGANISATION IN LIFE SYSTEMS.

It is messages from the outside environment, carrying information, that govern the forward movement of life processes. Societies have an analogue with life systems and resemble their components, human beings, in that they are temporary pockets of temporary decreasing entropy. A lack of distinctness or form indicates a high probability and therefore low information content. They are said to be boring or cliché ridden.

IN INFORMATION, HIGH PROBABILITY MEANS LOW INFORMATION CONTENT.

The relationship between life systems and the environment can only be studied and understood through the information messages that pass between them; and through the study of the communication facilities that belong to them. The future is going to depend increasingly on the development of these messages and communication facilities as a means of control; that is the messages between man and the environment; environment and man; environment and environment; man and man.

THE STUDY OF COMMUNICATION IS THE STUDY OF CONTROL THROUGH INFORMATION.

When communication takes place with another man, or machine or the environment, there is a message imparted to him or it; when the communication comes back the related messages contains information that is primarily accessible to him or it and not the original sender. The technique of communication does not differ if the intention of the original sender was to deliver a command or a fact. To know if a message is effective then the return message must be taken into account as an indication that the order has been understood or the fact received. An order must go out and a signal of compliance returned.

THUS ANY THEORY OF CONTROL, WHETHER IT GOES THROUGH A PERSON, A MACHINE, A BUILDING OR THE ENVIRONMENT INVOLVES THE THEORY OF MESSAGES OR INFORMATION.

The commands through which control is exercised depends on the amount of information imparted. The greater the information the greater the control. However, information is subject to disorganisation in transmission in the lack of coherence due to faulty structure and in any case can never be more coherent than the original sender. Control and communication are always fighting the tendency towards entropy and the destruction of meaning and the degrading of the organised, by feeding in new information either from new sensations or from memories, or from new commands made up from these which have been previously excluded from an information system, in order to decrease entropy.

CONTROL OVER THE ENVIRONMENT INCREASES AS THE INFORMATION IMPARTED TO IT GROWS, UNTIL IT BECOMES OVERLOADED.

There are limits of communication within and among individuals as there are with machines. Machines are limited by the way they are made and what they are made for. The human organism is like-wise limited. The world is seen through sense organs, and the incoming messages are subject to the limitations of the brain, the nervous system and the processes of collation, selection and storage which are very flexible but not particularly accurate especially in recall. These are very finite as are the possibilities of reaction through the limits of muscles and verbal reply. These limitations are partially dependent on the genetic inheritance of the organism; and partially dependent on what has been learned about survival in the environment, from direct experience and for humans on learned experience. All this is acquired at an early age and its very difficult to modify or deal with if the experiences are very different; making communication sometimes nearly impossible.

MISUNDERSTANDINGS ARE VERY LIKELY TO BE THE RESULTS OF MISCOMMUNICATION. A COMMUNICA-TION GAP.

An event can be considered to contain information irrespective of whether the information is important or correct.

UNCERTAINTY

Information is the reduction of uncertainty, that which reduces it is the signal or message. Uncertainty is not a measure of information, it exists before information and cannot co-exist.

INFORMATION REDUCES UNCERTAINTY.

UNCERTAINTY IS A MEASURE OF THE FREEDOM OF CHOICE WHEN SELECTING A MESSAGE.

Receiving a telegram does not reveal any information if the contents are already known. The probable or the improbable convey equal information. It is the prior range of probable distributions that determines how much.

REDUNDANCY

Messages sent over a communication channel are said to exhibit 'REDUNDANCY' if they contain less information than they could contain. The implication is that shorter messages, using the same range of possible symbols, could contain the same amount of information. Printed English contains perhaps 70% redundancy of letters than are needed for the information content; but it makes bad hand writing which misses and blurs letters, easier to read; or makes speed reading possible because redundancy is being filtered out.

Thus information theory provides a system of measurement dependent on a definition of the amount of information a message contains. The definition is such that the nature of the information is irrelevant. The concept is similar to that of the weight of an object, or the measurement of electricity as the potential difference across a wire. It measures the uncertainty before and after an event. Information can be considered as energy transfer. The concept also states that there is a maximum amount of information possible in any message, and this is dependent on the capacity of the communication channels. Whether in machines, or sensors or human brains.

FEEDBACK

Feedback is a phenomenon of being able to adjust future conduct on the basis of past performance; in which past experience can be used to regulate, not only specific movement but whole policies of behaviour. One aspect could be called conditioned reflex, another learning.

FEEDBACK IS LEARNING FROM MISTAKES

INPUT

A complex action is one which information introduced is called 'INPUT'; the selection of a particular incoming message at a given time is involved in a 'PROCESS'; and addition of this message to all other messages that have been selected and stores in the 'MEMORY' and obtains an effect on the environment called 'OUTPUT'.

MEN, MACHINES, ANIMALS ARE DEVICES WITH INPUT, OUTPUT AND PROCESSING POSSIBILITIES.

The control on the basis of actual performance rather than expected performance is known as feedback and involves sensory mechanisms that perform the function of 'MONITORS' and is one possible use. It is their function to control the tendency towards disorganisation; to reduce a temporary local reversal of the normal direction of entropy by energy transformation.

It is possible to regard the physical functioning of the living individual the operation of the newer communication machines, and the built environment, which is in itself a communication machine as precisely parallel in their respective analagous attempts to control entropy through feedback. They all have, or should have, sensory receptors at one stage in their cycle of operation. There should be a special apparatus, human or mechanical for collecting information from the environment and made available, in the operation of indivuals, machines or in a society reflected in the buildings as they are culturally determined. Buildings and the built environment should respond to individuals and social group demands that are put upon them.

To be valid and most effective in the outer world, the monitored performance must be the actual action and not the intended action, as is now studied and usually expected. It is this that should be reported back to the central regulatory apparatus. This has not been the habitual role in the analysis of society, anymore than it has been in the building, and has usually been ignored. For this reason, there are few results available of what actually happened, and about the effects, of a building as opposed to what was supposed to happen.

14

The Structure of the Message

LANGUAGE

The emergence of symbolic language, first spoken and then written, represents the sharpest break between animal and man. Many social animals have some sort of communication by signs and symbols, but spoken language is species-specific; the exclusive property of man.

The world is structured by the way the mind is structured. It would be expected that the study of language, as the most obvious means of communication, would reveal some basic concept of the way the mind works and is structured.

Many of the problems of "category" and "sequence of events" thinking have been caused by the assumption that because we utter one sound after another, or write one word after another along a single axis in time, that our thinking and perceptual processes must be actually structured in this way; in the form of a linear chain. However reading and hearing is not serial but is in a continuous code in much the same way as a track on a gramaphone record of a symphony seems to be a wavy continuous spiral curve under a magnifying glass, but is heard as fifty or more instruments playing at once. The restrictions on speaking or writing are caused by muscular and mechanical restrictions not on thought and thinking restrictions.

The unravelling of such linear coded sequences, or single variables in time is still little understood but can be represented by the concept of a multi-levelled hierarchy of processes made up of sub-assemblies. This requires a de-coding device, or a brain which has de-coding devices, to take the sequence in a linear form and retrieve

111

the pattern from it. Ideas are heard as a continuous stream of sound, like a space station reporting back to a tracking station. The machinery that generates speech is invisible and beyond the reach of inspection and introspection, working mainly in the unconscious. Psycholinguistics has shown that the only conceivable model to represent the generation of a sentence does not work from left to right but hierarchially; branching from the top downward as in the following schema:

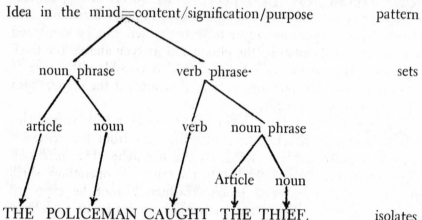

Idea in the mind=content/signification/purpose   pattern

noun phrase   verb phrase·   sets

article   noun   verb   noun phrase

Article   noun

THE   POLICEMAN CAUGHT   THE   THIEF.   isolates

The whole organisation of the sentence is the structure.

HIERARCHIES

At the apex of the hierarchy is the idea in the mind, the intention of saying something, but which is not yet verbally articulated; and at the other end is the individual and separate words. Used in this order their meaning is clear; used in any other order they would mean something quite different like 'The thief caught the policeman'; or they could mean nothing at all. It is essential to convey the meaning of an idea not only by using the right words but in the right order. The organisation is part of the meaning. They are isolates; the individual words; the sets; the noun and verb phrases; and the patterns; the order of the noun and verb phrases.

112

PARTS AND WHOLES

Each of these entities has two aspects. It is whole relative to its constituant parts and at the same time a part of a larger whole. It is a part and a whole. A servant and a master. It is a characteristic of all hierarchic systems that they are not chains of aggregations of elementary bits but are composed of sub-wholes branching into sub-sub-wholes. There are separate words which can be combined in a way that is dictated by the phrase and its type above; 'the thief' is possible, 'the caught' is not. 'The thief' is possible either side of the verb 'caught' but only one pattern is possible if the original idea is to be conveyed correctly.

This sentence has to be constructed to convey this particular idea and the recipient has to reconstruct the idea from the sentence. There is a movement up and down the hierarchy. The idea itself is not communicable by language. In the same way operations which generate language included processes which cannot be expressed by language. The attempts to analyse speech leaves us speechless.

The branching symbolises the step by step, hierarchic process of spelling out the implicit idea in explicit terms, converting the potentialities of an idea into actual motion patterns. It is in effect the sending of an order for some action to take place in response. The process can be compared to the development of an embryo as well as other facets of evolution. The fertilised egg; the idea, the hypothesis, the conjecture, contains all the potentialities of the future individuals, and these are 'spelled' out in successive stages of differentiation; each step is governed by fixed rules; 'the' cannot preceed a verb; but each has flexible strategies; it can come before 'thief' or 'policemen'; and all is guided by feedback; did this sentence get over my meaning to the recipient or will it have to be rephrased if it did not.

113

CONJECTURE, REFUTATION, FEEDBACK

The feedback is a trial and error; conjecture and refutation; at-tempt to convert an idea into sentences that can convey the original idea. The sap flows up and down the tree, in a search for the right word or the right sentence order. The fixed rules are the grammar, the organisation of the total discussion, the way the vocal cords work. There is also the variety of strategic choices; the selection and order-ing of the material, the words used and so on. This is the organisa-tional hierarchy or system. Running simultaneously through this construction is the idea of what way being conveyed, what the sen-tence signified. A sentence is a message that has to serve two masters, two interlocking systems at one time; one governed by meaning-signification and the second by meaning-structure.

HIERARCHY-SYSTEM

There is an imaginary story of two watchmakers who used two different ways of assembling the thousand pieces that went into making up a completed watch. One would put his watches together bit by bit, like a mosaic, and everytime he was disturbed he had to put down the total assembly and so it fell to pieces and he had to start all over again. Not very many watches got completed and so he gave up his shop and went to work for his rival in the town who seemed to be prospering very well, so that he could learn the secret of his success.

It seemed that his rival had found a different way of assembling the watches so that very little was lost by a disturbance. He was putting them together by making small sub-assemblies of ten parts which were complete on their own; and then putting ten of these together and so into a final ten that constituted the final watch. This method proved to have immense advantages. If he was disturbed then the most that could be lost was nine pieces instead of all the

114

elementary parts and what was more an immense amount of time was saved. If a disturbance came once every hundred assembling operations on average then it took the first man a thousand times longer to put a watch together; the difference between a day for one and eleven years for the other.

If these mechanical bits were translated into animo acids or protein molecules then the time scale becomes so astronomical that the whole life time of the planet would have only have been sufficient to produce an amoeba; unless the method of construction of complex assemblies from simple standard sub-assemblies is adopted as an explanation. Amongst all possible complex forms, hierarchies are the only ones that give the time that has been needed to evolve. Complex systems, can and will evolve from simple systems if there are stable intermediate systems that can be grouped together to make them.

LIFE SYSTEMS

Hierarchial systems are also far more resistant to repair, or damage or regulation as only one sub-assembly has to be removed or mended without destroying all the rest. It can be assumed that whenever there are complex systems, including life systems, then they will be organised hierarchially.

PARTS/WHOLES

The first universal characteristic of hierarchial systems is the relativity and ambiguity of the term 'part' and 'whole' of any sub-assembly. Wholes and parts do not exist anywhere in any absolute sense, including living organism and social organisations. Looking down a system towards a subordinate level it seems a self-contained whole; but looking up towards the apex then a dependant part. A section of troops in an army is a whole to itself but part of a company and so on up until the whole is called an army of which a

115

section is the smallest part. If systems are looked at as if they were pyramids then the whole could be described as pyramid with many smaller pyramids inside called parts of the whole but whole parts.

TRIGGERS

Turning a switch or pressing a button on a machine, although a very simple gesture in itself, releases extremely complex pre-set patterns. This mechanism of a relatively simple command, signal, order, message, is the means that an organism, or social group as well as a machine, is able to reap the benefits of the autonomous self-regulating character of its sub-units. These trigger mechanisms, setting out from the apex of the pyramid, at each step on the way down, releases pre-set action pattern which transforms the implicit message into explicit terms, from the general into the particular.

The process is the same in the production of articulate speech. The non-verbal inarticulate intent of conveying a message, triggers off the phrase-structuring mechanism, which in turn brings the rules of syntax into play, and so on down to the spelling of each phenome. It is the same process as the fertilization of an egg by another cell is sufficient to trigger off the incredibly complex hierarchial structure the human being; the egg contains all the rules for the manufacture of an adult.

OUTPUT, INPUT, FILTERS, SCANNERS

This is the 'output' the spelling out of intention into action. The opposite is the 'input,' the sensations and perceptions. The stream of information from the input comes as a mass of data that is then stripped of irrelevant detail, condensed, filtered, combined with other data and flows upward among the converging branches of the hierarchy. All the detailed workings of any army reach the General as a one page report. It is where the buck has to stop and where a new trigger mechanism has to be started out. On the motor side are the triggers, and on the perceptual side are the 'scanners' and 'filters' through which the input must travel. Their function is to analyse, de-code, classify and abstract information, until a multitude of sensations, which constantly bombard the senses, are transformed into meaningful messages. These filters are for the most part unconscious, depending on the activity on hand, the mood of the moment, likes and dislikes, programmed from childhood.

A sentence as a whole is a Gestalt image, giving meaning or the idea that the sender wishes to convey, but the parts are made of phrases and phonemes of words.

Systems are divisible into sub-assemblies, the general patterns; sub-sub assemblies, the sets; and individual mechanistic units, isolates. The number of levels downwards is called its 'depth' and the number of isolates, the 'span.'

Each pattern and set is governed by fixed 'rules,' the way they are put together, and flexible 'strategies,' the choice of which way to develop; the way it connects with others to form some particular purpose or meaning.

STRUCTURE/ORGANISATION/CONTENT/FUNCTION

Thus it is possible to distinguish between two different kinds of systems; a 'structural' system that emphasises the spatial aspect; the anatomy, the topology, the organisation, of a system, and a 'functional' system which emphasises the progress in time; the purpose, the idea, the content. Space and time cannot be separated and so structure and function cannot be either; they represent complimentary aspects of an indivisable spatial-temporal process; but it is often useful to focus attention on one or the other. The 'part' within 'part' character is more easily recognised in the 'structure' than the 'functional' systems.

The army is structured into officer, non-commissioned officers and men, soldiers, so that 'functional' messages, or orders, can be disseminated quickly without the general having to tell every soldier himself. The advantage of sub-assemblies over elementary parts. The whole purpose of the army system is to fight wars and this purpose is beyond question by the army because that is why they exist; to abolish the idea of war is to abolish themselves. By Gödels Theorum the axiom must be assumed before a system can be logically constructed; to remove the axiom is to make the system illogical. Given that there are wars there must be armies. So whether it is for language, or music, or mathematics, or armies they have functional and structural systems linked together to describe them.

MEANING

If an order is to be effective it must have a 'functional' purpose and it must have a 'structure' hierarchy to pass down; at the same time the order itself must have both 'functional' and 'structural' content then the connection can be seen with information theory and the two different uses of the word meaning; that is meaning . . . it *signification*, the function system of the order and meaning . . . the way it is *organised*.

In functional hierarchies the input intention is particularised and spelled out in its descent to the base; in the perceptual hierarchy there is the opposite effect. The output hierarchy concretises; the input abstracts. The former operates by trigger mechanisms, the latter by filtering or scanning devices. Triggers release complex outputs, scanners function the opposite way; the former by simple coded signals, the later by the conversion of complex inputs into simple coded signals.

The principal of self-regulation is in fact fundamental to the concept of systems. To function, a semi-autonomous sub-assembly must be equipped with self-regulatory devices. Its operation must be guided, by its own fixed cannon of rules, but also by the variable experiences of input from the environment. There must be a constant flow of information concerning the progress of the operation back to the center which controls it; and the controlling center must constantly adjust the course of the operation according to the information fed back to it. This is the principal of feedback control. The coupling of the output to the input. Feedback from the environment guides and stabilizes a pre-existing pattern of behaviour.

The environment is influential on the flexibility and rigidity of behaviour. If a skill is practised in the same unvarying conditions it tends to degenerate . . . into stereotyped routine, and its degrees of freedom solidify. Monotony accelerates the enslavement of habit; mechanisation is said to spread up the hierachy or the functional system becomes more subordinate to the structural system. A changing environment demands flexible behaviour and the process is reversed and the functional system takes over. The skilled car driver can let the details of driving become routine and subconscious until an emergency arises and new forms of flexible behaviour are suddenly needed.

PATTERNS, SETS, ISOLATES

It can be said about systems that on successively higher levels there are found more complex flexible and less predictable patterns of behaviour; while on successively lower levels there are found more and more mechanised, stereotyped and predictable patterns.

All skills tend with increasing practice to become more mechanised routines. Monotonous information facilitates enslavement to habit while unexpected information contingencies reverse the trend and may result in new improvisations and innovations.

Higher areas of a system only communicate with lower ones in order and through the 'proper channels'. The short circuiting of this rule causes disorder of various kinds and possible breakdown of the system.

The hierarchic genetic lesson is clear; the simple sum of separate events, molecular, social or otherwise, is insufficient to represent a living whole. Some property is lost, when cells are broken apart for analysis, they are 'killed,' although when they are all together they exhibited a type of behaviour which is described as being alive. No matter how many ingredients they may contain, the bio-chemist's test tube mixtures are dead, but when these same ingredients are organised in the structure of a cell they comprise a system that is alive.

In any living system to remove a part does not just leave an empty space, it kills all that part of a system that is dependent on

119

it, for which it is responsible. If the system cannot reform quickly it dies, or if it does not reform it becomes something very different. The whole is greater than the parts. The unique properties of a complex system are not explicable solely by the properties that can be observed in its isolated parts. To study the isolated parts does not explain the collective interactions.

There is a conflict between the two approaches to the theory of life and the organisation of living systems; one suggests that the answer lies in separable chemical reactions, the other that the uniqueness is a property of the whole cell and arises out of the complex interactions of the separate events.

The inability to see such living systems as interacting has led to some temporary successes at the cost of the whole. Contamination of water or the air, pollution of cities, creation of deserts and dust bowls have given short term pleasure at the expense of the future.

15

The Meaning in the Message

INFORMATION

Information is a measure of the freedom of choice in selecting a message. The greater the choice the greater the uncertainty that the message selected is a particular one. Thus greater freedom of choice, greater uncertainty and greater information all go hand in hand.

Information always refers to the future, and is therefore predictive, with a degree of probability defined by the number of alternatives controlling the possible result.

Information carried by a set of messages is a measure of its organisation. The purpose of a message is to convey information; messages come from everywhere; from other people and other animals and from the environment.

CONTENT/SIGNIFICATION/ORGANISATION/ STRUCTURE

THERE ARE TWO PARTS TO EVERY MESSAGE:
WHAT IT MEANS:—ITS CONTENT, ITS SIGNIFICA-
TION
WHAT IT MEANS:—ITS ORGANISATION, ITS STRUC-
TURE

121

A message that is meaningful to us is conveyed by a pattern which is recognized by its organisation. The more recognisable the organisation the easier it is to understand the meaning; but the less information it contains; it is said to be a cliché. To organise some pattern is to reverse entropy; the tendency towards entropy or disorganisation, we say that we have sent a message. The more complex the organization, the less probability of "yes" or "no" choices, and the greater chance of a message getting through.

It is the function of poets that they send messages of such low probability that they cannot be easily misunderstood; on the other hand it is the function of politicians to send messages of such high probability that they cannot be understood and therefore upset anybody; nothing can be done with their clichés and so they are virtually informationless. This is the reason why the application of these clichés to controlling the cultural or social machine proves so ineffective. Control is related to the amount of information received.

Meaning connotes significance but also 'purpose,' 'intention,' 'design.' Such things only exist in the emitter of the communication, and the receptor insofar as he or it is able to put himself in the emitter's place, assume his role and guess what are his real intentions. It can never be in the transmission, the sign, which clearly contains no intention in itself.

KNOWLEDGE

SIGNIFICATION CONTAINS INFORMATION ABOUT KNOWLEDGE FROM LIFE EXPERIENCE. STRUCTURE CONTAINS INFORMATION ABOUT KNOWLEDGE FROM LEARNT AND STORED EXPERIENCE.

MEANING

There are two aspects of the word meaning in information theory;

MEANING AS SIGNIFICATION
MEANING AS STRUCTURE.

The first is concerned with semantic meaning; the word 'walk' signifies activity of the legs.

The second is concerned with the structural meaning in the English language; it is a verb.

SEMANTICS

Semantics defines the extent of meaning and controls its loss in a communication system;

Semantics is concerned with meaning;

Semantic reception demands memory;

Semantic receiving apparatus neither receives or translates a language word by word, but idea by idea, and very often in an even more general way.

The sentence 'I walked around the house' has meaning because it has internal structure and can be measured. But if someone was to say 'I cammeled around the house,' then it has no meaning, although each word in itself means something. The words have lost their meaning because the structure has gone.

A trained musician has a greater understanding of musical structure, he can interpret the structure where the untrained would find that there was a blockage in the brain that would not allow this information to pass. The musician can appreciate the structure-organisation as it would exhibit to him a significant form of pattern and further understanding.

AESTHETICS

The appreciation and discussion of this internal organisation is normally called AESTHETICS. The question when discussing or considering a sensory image is about how much internal organisation does it have and how well are the interactions interrelated. The activity is concerned in finding out how many things rushing around in my head can be related to this image; how much learned and stored experience relates to this new experience. The more that can be found the greater the pleasure, the more that is known the more that can be found.

AESTHETICS IS TO PERCEIVE THE ORDER AND STRUC-TURE IN THINGS THE MORE THAT IS PERCEIVED THE GREATER THE PLEASURE. THE GREATER THE KNOWL-EDGE THE MORE THAT IS PERCEIVED. TO SAY SOME-THING HAS MEANING IS SAYING: THAT IT IS STRUC-TURED, IT HAS SYMBOLIC CONTENT OVER AND ABOVE ITS STRUCTURE.

Marshall McLuhans great mistake is to fail to differentiate be-tween the two different uses of the word 'meaning'; he has not been able to distinguish between the two.

It is possible to look at modern social and cultural experience through information theory, very precisely, and without having to get involved in language or the effects of its structure, like the alphabet.

What is happening is that the meaning-structure is decreasing and the meaning-signification is increasing; but as the amount of meaning or information in a message is finite, there is now in our culture less structure and therefore greater amounts of signfication. The questions are more about why and what and less about how.

Change is a situation where the content is much higher than the structure. There is a great deal to think about and very little ap-parent organisation to structure it around. We are now moving from a period where structure has dominated towards one where content is becoming more important as a source of information.

Marshall McLuhan's thesis runs as follows; there is a new world of sensibility being created under the aegis of the new electronic media. Pre-literate and tribal man lived in a rich aural-oral world, structured by myth and ritual, its modes of awareness being tactile and auditory, its values communal and sacred. The Gutenberg revo-lution exploded the world of the tribal man, creating via print the open society, modern individualism, privacy, specialisation, mechani-cal repeatable techniques; all at the cost of cutting off from richer auditory experience.

124

Hence fragmented-specialised impoverished Western man is Gutenberg man, a necessary victim of the visual emphasis given by print technology. The electronic revolution however once more makes the aural-oral experience central, promising liberation from print, demanding participation rather than print based passivity, and restoring us whole and in harmony in a reconstituted tribal society throughout the global village. The electric, instant-information, media are going to re-tribalise everyone where they like it or not, and irrespective of their content.

The idea of any role that content may play in a communication is rejected. The fact that his own work has content is an irony that passes him by. His message, his content, comes over on a very large scale indeed judging by the response that he gets using a medium that he claims is impoverishing and detribalising, and that his most avid followers are the re-tribalised. He totally ignores the power of ideas, ideologies, values, accumalated wisdom that is necessary for basic survival, to say nothing of the facts of geography, econonic development, politics, and different varying cultures all over the world.

Deeply enshrined in the thesis are all the central 19th century dogmas. The mechanical effects of electric media can causally determine behaviour; that technology creates the people to go with it, much in the same way that birds fly because they have wings; that there is a distinction between the literate arts and rational sciences and technology, a constant battle that means one or the other must win. The only alternatives are between the tribal man all feeling and all sensing and the print abused man feeling and sensing nothing. A millenium and utopia for all in the grand manner.

Certainly there is an obvious craving in Western society for a more richly orchestrated life of the senses, for aural-oral modes of sensory experience and communication with others as a revolt against passive consumer oriented roles. The idea of full time creativity for everyone, to process rather than product, to getting involved, to acting out is possible, but the question is much more concerned with whether it is probable, or necessary, for everyone or only those who need it and have so far been deprived because of the bias of their culture in one direction instead of another. If the mechanical world was a result of print technology who is going to run and maintain that technology when all the books are gone and the return to the tribal man to be complete?

The medium is only one part of the message, the other part concerns its content. A message that is totally concerned with its medium has no message at all, as can easily be demonstrated by turning a television set on after the transmitter has closed down or a technical fault eliminated the voice and the pattern or picture.

There is a direct law about the amount of total structure and signification that a message can carry. The more of one the less of the other. This is because the brain is a limited machine and meaning is not a real thing but only the amount of information that can be got out and that depends on how much has been put in through education and practical experience. Information is a natural phenomenon like electricity, and the brain an information processing plant with a limited channel capacity, so the messages must be limited also.

Structure is relative and depends on the way any culture views the organisation of their environment at any time. The difference between the operas of Mozart and Wagner when comparing them for information content is that Wagner threw away conventional views of structure and rules and so was able to increase the signification or relations between things, but Mozart stayed with the structure of his time reducing the information content.

The composer Stockhausen has almost completely eliminated structure from his work but not meaning which has become very information rich in signification. Marshall McLuhan is making a meaningless statement when he says that meaning has been thrown away.

What he really means is that we now have more signification and use less structure.

What all this signifies is that our present western societies are going through a phase when more value is being put on what our personal and social games signify rather than the way they are structured. The games we play are changing.

The pendulum is swinging back from a period when the major concern was with organization; rather than the signification which

126

was dominant in the Medaeval period. Since then the world was assumed to be highly structured, that is rational and open to dissection, and the results in terms of investigations into the way things appeared to be organised advanced very fast.

But structure is in the precept not the object. Greater and greater amounts of structure produced greater and greater amounts of signification at the same time and the prevailing attitudes of ignoring or being ignorant of it does not make it go away, and therefore there was a great deal more of total meaning available.

So what is happening at the present time is that people are becoming aware of this immense store of information lying half hidden around that needs organising in some way. But first this information and what it signifies has to be understood and this is impossible to do by people who are only acquainted with the procedures of the manipulation of organisation, so new people must be trained or train themselves to understand this new information and then set about structuring again and so on.

This can be seen to be happening in the arts with the search for literal structures and in the sciences with the search for patterns. But this is in the most advanced fields, in less developed areas there is still a search to understand the meaning, signification, of it all. The pendulum swinging backward and forward for each person and each society and each culture at a relatively different time in space.

This is the meaning of the present western cultural revolution; there is nothing absolute or utopian about it, and it is totally relative in time and space.

The more useful and operative explanation of Marshall McLuhan's analysis of present cultural problems is a follows;

There is a new culture, more advanced than our present culture, developing in some areas as a result of an increase of information created by advanced electronic information-organising tools. This has demanded an complementary increase in the understanding of the significance of this new information.

To understand this new information it is necessary to seek out in the first instance, and train, as soon as the educational system can be re-directed, people whose mental abilities and experiences are closer to the spatial-mathematical-mechanical end of the spectrum than the verbal-numerate end. There is a cultural lag in grasping the need for this re-direction in the education system which is still almost totally verbally-numerate oriented, and this is causing a potential violent situation resulting from the generally unconscious realisation that the change is required but is being frustrated.

The change does not require the disappearance of all verbal-numerate people, who must continue to organise and run a society whose technology is not going to disappear in the period of transition. The need is for the education system to reflect the new interest in the signification rather than the organisation of the society which now interest many students.

VERBAL ABILITIES

There is a connection; Verbal ability and personalities concerned with meaning-structure as information handled by classification, specialisation individuality, passivity, detail and in this sense could said to be 'de-tribalised.'

SPATIAL ABILITIES

There is a connection between: Spatial ability and personalities concerned with meaning-signification as information handled by total view, exploration, co-operation, myth and ritual, and in this sense could be said to be 'tribalised."

The cultural needs of the present and the near future are swinging the emphasis of enquiry away from the verbal towards the spatial side of the spectrum.

FROM A STATE WHERE THERE IS A SWING FROM STRUCTURE TO SIGNIFICATION: FROM ORDER TO DISORDER: TO A STATE OF CONTINUOUS CHANGE.

It has been discovered by experience that a sequence of musical

128

notes can produce immense ranges of emotional feelings. Wagner, for instance, found that he could achieve the greatest effect by allowing himself the maximum possibilities by removing as much structure from his music as he could. The result is an emotional experience which has been labelled extremely 'romantic'. In effect the lack of structure allowed a great deal of signification; as with a poet the message was made more easily understood because the amounts of bits of information being received are large.

Mozart, on the other hand, put in a great deal more structure than signification and therefore there was less room for emotional appeal but more for the appreciation of the structure and organisation in the mind. The result is an experience that has been labelled classic'.

The description of 'romantic' is given to messages that deliberately appeal to a desire for emotional stimulation by signification; and 'classic' to messages that appeal to the desire for emotional stimulation by structure.

It was much less easy for Mozart to write a meaningful love duet than it was for Wagner because Wagner's greater emphasis on a mass of loosely structured information made it easier to understand his message than Mozart. Mozart's elaborate structure left little room for anything else and so very often without prior knowledge it is possible to confuse the comedy arias with the love arias. For the same reason Bach or Pop music or musak can be used for background music because the demand on the mind to sort out the meaning-signification rather than the organisation of the message is not so great. Structure is less involving than signification.

The amount of structure also has a bearing on the amount that can be remembered; the more structure to any message the easier it is to remember. The work of Stockhausen is almost impossible to remember because of the lack of structure. The clue to learning something which has little structure is known by students who compose a jingle or nemonic to help recall.

'Happenings' are events or experiences that are composed mainly of signification with little structure and so are difficult to re-construct or remember. But it is clear that most people cannot exist for very long without needing to structure what they are doing and that total or near anarchy is not possible. The 'hippy' movement began to die as soon as it became obvious that beads and bells and long hair were becoming a uniform, that is a structuring device, at the expense of the spontaneity of the movement. Every kind of happening soon acquires a structure and new happenings have to be created, to replace them. And so all events move across towards greater and greater structure and ossification.

There is no way of knowing exactly how this is going to affect events in the future. Clearly the perception of structure for some people is highly enjoyable; appreciation of structure is an aesthetic experience; organising a structure is both necessary and highly delightful. On the other hand a lack of structure in life leads to feelings of aimlessness. Drug taking and dropping out are searches for structure and meaning in life, in the same way as work. Enough money provides enough structure to look for further signification. Restructuring or de-structuring is usually to be able to structure another way that allows the possibility of more signification content.

It is reasonable to assume that present art movements, for instance, are a swing of the pendulum towards signification while a new structure is being worked out, rather than a series of events that are leading away from structure altogether. It can be assumed that the pendulum will back a period of consolidation before another withdrawal to leap forward.

Even the learning process that proceeds from blind formal copying at first without understanding the signification but only the form, structure, is in itself a structuring for those who do not yet grasp the significance of what they are doing. After they go on to question informally and then technically; finally breaking away, hopefully into the possibility of new awareness-signification and away from the old formalism structure; only to provide a new structure for their successors.

If artists always seem ahead of their time it is because they are seen to be providing the current period with material of signification and new structure so that yesterday's art becomes today's structure. But it is a mistake to think that new structure will resemble the old. They reflect the way that things are being looked at and this can be quite different. It takes some time to discover all the restraints

130

of any new structure but once this has happened then there is a rejection and a search for a new one. This is the procedure of innovation and consolidation and at the present time seems to be speeding up in time.

Structure is now thought to be about relations and not probabilities or equal and opposite forces as in the last periods of thought about structure and organisation. The working out of this idea is causing a great pressure for change. The next few decades could be described as a state of continuous change.

Section IV

The Nature of Design

Designing buildings, or towns can be seen as the selection and organisation of systems embodying cultural information for a particular end and as making proposals for meeting that end assuming that it is but a tentative solution. Any built product that reflects this must acknowledge it by being open to criticism and change as soon as it ceases to correspond with the desired information systems. In this way buildings must be seen as continuous approximations to a desired end and must be able to adapt, as living organisma adapt to change in incoming information. This is what is meant by flexibility.

What is needed is a social architecture where arrangement among people evoke what they really want to see in themselves. Instead of looking to a professional elite for the solution of social problems we should look to the very population that has the problem. People who have problems know more what must be done. We have to learn to co-operate in a way that evokes intelligence and help others to invent their futures as we expect to invent ours.

Buildings are models of the future; or they should be. They are involved in information transmission, the communication of ideas about our future relationship to the environment; the way it is perceived now and the way it is perceived when it is constructed. To do this models have to be constructed that can be run faster to see what the outcome might be with the information now available as a starting point. These advanced information models have to be left open to adapt if the future is open and changing.

135

In effect, it is playing a real life game where the rules, or constraints are laid down, but the strategies for action are very open and flexible. Up to now designers have been much more interested in changing the rules rather than their strategies. The resulting effect on the ecology and the built environment has been disastrous and cannot continue much longer. Design or problem solving can be learnt through the playing of games in the same way children learn about the real world situation by playing of games also.

The success of such models depends on the amount of information available to the designer. Architects up to now have been relying on very poor information and obsolete concepts from times that have passed long ago, when available information was very much less. If architects are ever to become viable again they must become aware of the most advanced information now available and how this is affecting the evolution of their own culture or any other culture that they might work in.

ARCHITECTURE IS DEFINED AT THE PROVISION OF SUITABLE TOTAL ENVIRONMENTS FOR THE EXCHANGE OF EVOLVING CULTURAL INFORMATION.

16

Architecture as Example

COMMUNICATION

If architecture is looked on as the provision of suitable environments for exchange of cultural information, that is that it is concerned with communication, it is important to know how to put in and take out the meaning from such a message system. In a broad sense the discovery of architectural meaning could result from an examination of its mode of production and creation both now and in the past.

ARCHITECTURE

Architecture in a sense not restricted to the work of 'architects' but the whole of the built and modified environment can be seen as a record of past cultural events and beliefs and therefore as a communication between the past and the present.

MEANING/SIGNIFICATION/STRUCTURE

But there two elements of 'meaning'; the 'signification' or functional message system; and the 'structure' or organisation of that message, the structural message system; there can only be a limited

137

amount of both added together and if one is extensive there is little room for the other.

In other words; within any fixed system of symbols or events, structural meaning is a prerequisite of signification meaning. Unless correlation exists between one symbol system and another symbol system or a system of real events; between langugae and a real event there can be no external signification. Unless the symbols themselves are correlated there can be no special rules whereby the internal signification can be learnt; so unless there is a grammar there cannot be a language.

DESIGN

Thus the following considerations in the position of architectural theory is concerned on the one hand with what architecture should be concerned in signifying but mainly with the procedures of its production and creation; that is its 'design'; its structural meaning— which is what information theory studies.

In order to have signification there must be internal structure because nothing can be signified unless there are rules or knowledge of how it is signified. In the case of Stockhausen's music the inability to remember it came not from lack of quality or some other judgment on its significance but from its lack of structure. If a building is to be remembered then its structure must be sought so that it can be re-constructed. Structure here means the structure of its message not its structure in terms of its construction. The more obsious its structure the easier it is to remember but in consequence, of course, much less is its significance as a message system containing information.

It is of enormous 'significance' that discussions about contemporary theories of architecture are almost impossible because there is no structure currently available for such discussion to take place. It is no coincidence that at the moment there seems to be a search for 'Design Methods or Procedures' or other structuring theories which include "image making."

As defined in the theory of communication of information, architecture has no relation to real 'functionalism.' In architecture it has always been about 'how it does it' not about 'what it does'; problems have become questions of how to express "structure" without having to cover it with plaster or paint; whether all the mechanical services should be left exposed to express their true nature; whether basic materials should be left exposed as expression of the nature of the material.

In all this the actual purpose of the building never changed in the architect's mind. The cultural purpose was to express the architect's own knowledge and experience and ability to use the concepts that he had inherited from the classic traditions; a view of world order enshrined in his position in society and his professional Institute and his audience's adulation for his individual genius.

When other areas of the culture were concerned with the significance of the new information, the function of the culture and how it could be be structured, architects were still applying their highly structured concepts, and inevitable the "design methods," unquestioningly, but by trying to rearrange the mechanical structures they started to introduce some of the new information and the incompatabilities began to grow and compound chaos. The "design methods" used obviously depended on the analysis of existing buildings that were the result of solving social and cultural problems that no longer existed; the method of Analysis-Synthesis-Communication was obviously based on the philosophical process known as 'induction' that had been discredited as a means of problem investigation for some time; the method of designing with paper and T-square was so structured that there was no possibility of changing it without having to reject the whole of the basic premises of architectural theory, processes and professional structure; the whole teaching, competition and assessment is founded on the production of drawings, differ little or at all from the plans, elevations and sections of the Renaissance 'Masters.'

Perhaps the climax of this dilemma has now arrived with the Competition for new school and club building for the Architectural Association in London. Competitors have been asked to submit proposals for a building to house an advanced course in teaching architecture with the very welcome proviso that a Competitor can submit a scheme that corresponds to his ideas of a teaching course if he does not agree with those set out in the regulations. Careful reading of these conditions reveal, however, the appalling naiveté that is current architectural theory.

The purpose of the building, the educational aims, makes no attempt to define what architecture is thought to be about, but the assumption has to be that nothing is thought to have changed from traditional theories because there follows a long eulogy to the cult of individualism with students doing individual projects of their own choice, with a "delicate balance between freedom and guidance between student and master," good old craft apprenticeship under another name. There is no mention even by implication of current real problems and their complexity that needs highly organised teams with highly sophisticated techniques for their solution.

It is not even possible that this message could be changed as it is completely tied to the dual requirements of the examination of a highly structured professional Institute and the University, which in themselves cannot allow any great modification of traditional practices. Even within these restrictions it might have been possible to alter the organisation to meet new demands but there can only be found lists of requirements for private offices, specialist departments, specialists, craft workshops, private research, lectures, 'criticism' and the stock in trade the central library, the fountain of all verbal-literate culture elitism; which is to be the center of the learning process; in fact here is all the paraphenalia of the bureaucratic teaching whose crudities have brought out many universities into a state of revolt. There is no hint anywhere that architects are preparing to face up to the realities of their time and to try and train people to meet the challenge. The tragedy is that the A.A. is looked on by many as being in the avante-guarde of architectural theory and a guide line to the future of architectural education.

Perhaps the final indignity is that any competitor that wishes to disagree with the ideas as set out in the regulations is invited to dissent as much as he likes providing his alternative is submitted on "plans," "sections," and "elevations" like everyone else. You are allowed any colour as long as it is black.

The object of illustrating this dilemma at such great length is to sort out exactly where things have gone wrong. Architectural history shows great variation in the content of its messages, showing large shifts from signification to structure and back. Gothic cathedrals with their large open obstructed spaces allowed the full might of the Mother Church to display itself. The spaces showed little complex organisation in comparison to the signification of the Church. In consequence the builders, in order to achieve these spaces, developed the efficiency of their materials and mechanical-structural techniques to their uttermost.

The Renaissance, on the other hand, veered to the other extreme; the spaces in their buildings became structured and organised in a very complex way as a reflection of the complexity and organisation of their culture. Apart from some large domes and arches building techniques were hardly needed to be much more advanced than that of the Romans and certainly very little was taken over from the Gothic period that preceded it.

In this way it can be said that the built environment is a cultural message system in which the purpose is conveyed, not by the building fabric, but by the spaces in the behavioural sense are organised. Function is a measure of the amount of cultural signification not the way it is built. The way that a building is built is irrelevant to its function.

FUNCTION

The significance contained in the new information has altered immensely and in consequence the old means of structuring it are proving useless, but the new attempts to find a system of structuring in all fields of knowledge, are still very vague, incomplete and experimental so the tendency is to leave every thing very open and unstructured or organised rather than solidify things too soon and find a mistake has been made that will distort investigations in the future.

The 'function' in this case of the new built environment as a reflector of contemporary culture, is to correspond to this situation by remaining also very unstructured. It is here that the fundamental theoretical mistake has been made, because the purpose and organisation of the cultural message has been confused with the way that the medium that carries the message has been made. This has lead to total theoretical confusion and subsequent theoretical chaos that has been left to individual whim to sort out, or an about-face to the 19th century again for those who cannot face up to it.

SIGNIFICATION, STRUCTURE, CONSTRUCTION

Confusion in architectural theory has arisen because of the confusion between three different aspects of the communication of a message:

—The signification of the message; the 'function' or 'purpose' of the message which is information about the culture.
—The structure of the message: The organisation of the message to convey the signification. The greater the organisation the less signification that can be carried, therefore the less information about function or purpose.
—The medium that carries the message: The mechanical means that carries the message from A to B; sound waves and the mechanical means to extend them; or light rays; or buildings and constructions:

A message can be sent by any number of different media, a telegram, a telephone call, a small boy, or by a fixed system of hand signals; what is restricting in these systems is the small number of messages that can be carried at any one time. A face to face confrontation between people is more involving because many different message systems are operating at the same time, only one of which is the voice, although even this can carry increased information by using inflection and so on. By all these different means so much information is being received that the purpose of the message is usually very clear; so much so that the behavioral messages from the body may contradict what the voice is trying to convey.

A great love affair is easier ended on a telephone than in a confrontation because of the telephone's very limited capacity to convey messages. The greater the number of messages the greater the probability of understanding all the ramifications and possibilities of a piece of information. Television performs this task better than a telephone for this reason; there are more pieces of information being transferred at any one time; so it can be said to be more involving. But it is not the medium that is more involving, it is the number of messages.

But none of this should be confused with the amount of information that a message carries as this depends on the way the message is constructed. A primitive telegraphic system using a rigid Morse code to structure its messages was very slow, because of the linear sequence of dots and dashes limitation; and very low in information content because of the way the message was cut short and abbreviated by the structure of its language; mistakes were very frequent. When person to person conversations could take place the use of the voice allowed the possibility of more information and less mistakes; but it is still very easy to misunderstand especially in very emotional charged situations and mistakes are still often made. But none of this has anything to do with the wires and relays except in terms of mechanical efficiency in transmitting and receiving the electrical signals that carry the message.

What architects have been concerned with expressing is the 'wires and relays' of their buildings and artifacts not the message and its 'function' or 'purpose' which is information about the prevailing culture. Questions of expressing the structure or the materials or the lift-shafts are the equivalent of expressing the internal workings of a headset or the wires of the instrument itself and its connecting wires and exchange to the others.

This is not 'functionalism' but 'expressionism' or 'formalism.' A concern with the form of the message carrying system and not the function of the message. In terms of information theory 'functionalism' is information about the purpose of any part of the built environment, and is part of a message which is structured by the allocation of spaces, in the behavioural sense, that correspond to the use that is approved by current social mores. As far as the mechanical-structural system that conveys this message is concerned the main objective should be to work as efficiently as possible so that it interferes with the message the least. It has no message in itself at all only the capacity to destroy the message by working inefficiently. Attempts to interfere with this efficiency for whatever reason, whether by lack of skill or knowledge, or for other reasons, only garbles the message of the building, as is only too obvious when looking around the present urban scene.

Education can take place in any building whatever it is made of providing that the spaces are appropriate to the way teaching is organised in that particular society. For this reason old buildings can be used for different organisations in following cultures providing that the spaces were suitable. If the message was in the building fabric, the medium, it would destroy or disrupt the new enterprise. This clearly does not happen.

A school building made up of a row of similar shared classrooms is often said to be boring. The analysis of this situation from the point of view of information theory shows that the reason why is that the educational process has become so rigid and over-organised in its content that there is very little room for signification; the purpose has become boring so the spaces are boring; high rise apartment blocks are boring because their overstructured spaces allow no possibility for human waywardness or even normal behaviour like a mother being in close proximity to her child.

Conversely modern churches, or factories, or super-markets are totally unstructured so that anything can take place in them. Signification has full license to do what it likes and to change when it wants; so production can change or cars be sold where the week before there was a food market. On the other hand people do not know what the building is 'for' they do not read its signification from its spatial organisation.

Mies van de Rohe made spaces into which almost anything could go and which was often made to accommodate hopelessly incompatible activities like laboratories and drafting rooms. His main preoccupation however was in the expression of the way a building was built. Attempts to emphasise certain areas of the spatial-mechanical systems very often made absolute nonsense of other parts to the detriment of climate control and working conditions. He made the building inefficient to satisfy personal aims. He did not deny the organisational mores of his culture, taking the 'classic' position of passing messages of great organisation and therefore little information content according to tradition; but playing with the images of technology rather than the concepts and methods of technology, treating it as an art and not an applied science; that is a way of thinking about problems. Re-emphasising the belief that art is a more worth while occupation for the elite of 19th century culture. Mies was never a 20th century architect.

Le Corbusier seems to have taken converse attitude with spaces, prefering to organise them, but took the same line as Mies interfering with the building components to fit the images that he acquired through more primitive building types totally ignoring their cultural purpose in the original setting.

The Beaux Arts rules of design underly both these architects and their followers; what was and still is taken for their modernity of thought was not a radical reassessment of the meaning of their culture as was happening in other fields, but the rejection of the previous rules, which were quite arbitrarily based on classical theories of the way the fabric of a building should look, in favour of images taken from something that seemed more modern or more individual.

MODERN MOVEMENT IN ARCHITECTURE

Everyone has been gravely misled by the Modern Movement in Architecture into believing that it was an equivalent intellectual movement to the revolutionary theories and work being carried out in music, painting, sculpture at the turn of the century and that in fact some of the fundamental changes had been made in the practice and procedures of architectural design.

DESIGN METHODS

But any investigation behind the image making on the facade will show that very little if anything had been changed since the culmination of the Beaux Arts theory of design at the end of the 19th century. This Design Method developed from Vitruvius Alberti? Palladio became very highly developed and methodical, providing a fixed scale of values, a strict scale and architectural vocabulary and grammar that were believed to be fixed in time and place and so suitable for anywhere on the earths surface. It provided not only a fixed Style that everyone could immediately recognise, but also served as a vehicle for learning, both for the practitioner and the knowing educated viewer, and was internally consistant. The system was a closed one, based on a number of limited ideas, that were mainly useful for administrative convenience, for making awards and critising performance, and gave scope for the clever from making bad mistakes. It showed an unshakable belief in a future utopia consisting of universal application of its rules and was completely isolated from contemporary events and immune to social and economic change and the advance of technological invention.

It is difficult to see how the Modern Movement has deflected in any way from these same desired ends.

A close examination of the subject matter of writings and theorisings on Architecture up to the present time shows that they deal primarily with post-facto products of "good" or "bad" design orientations. The question has never been over the significance and meaning of what is good or bad, that has been taken for granted, but always about the best means to achieve the required result; how to set out a procedure for design that has all the advantages of the Beaux Arts system.

Implicit in any design procedure put forward, whether by those investigating the construction of a universally applicable "Design Method," or "Corbusian Method," or "Mies van de Rohe" or any other 'Method' so far, is the idea of a specific required solution based on values that have not been questioned. These required solutions have the effect of not only ordering the solutions but the problems as well. These required solutions have come not out of any incompatibility of human behaviour and the environment, but out of an Architect's personal problems, and predelictions for the 'expression' of personal values that are not valid or significant.

The current predominant question of architectural theory is the

146

search for those structuring theories that could give a lead to the direction to the future of the profession, and more particularly still its education system. But there is extreme doubt as to whether this search is in the right direction.

The wrong path seems to have been taken at least 60-70 years ago when theorists failed to appreciate that more was changing in their culture than visual appearances. There seems to have been a total lack of awareness, that still to a very large extent exists, that the culture was taking on a very different significance and complexity that could no longer be structured with previous organisations and institutions.

In the face of all the new information, new techniques for handling and structuring it, were being experimented with throughout the arts, in music, painting, literature, sculpture, but not in architecture which still only copied the visual images second hand from the mechanical technology, ships, blast furnaces, green houses; or from the visual arts, cubism, constructivism, futurism; many of the practitioners were themselves only interested in dressing up 19th century social and cultural values in new clothes; very few even being in the smallest concerned with major social revolutions that were taking place in Russia or less successfully in Germany after the First World War.

It was this desire to stay behind in the dying mainstream of the 19th century whose values and concepts stretched back through the age of mechanical revolution, and the literate-categorising-individual eliteism of bureaucratic nationalism; that has given rise to present problems. The seat of the trouble is that theorists, realising that they must come to terms with present day problems, have tried to do so with the wrong concepts, they have tried to solve 20th century problems with 19th century techniques.

This is not surprising as no one has ever raised doubts about the matter as they have all been wedded to the same objectives, which have been consolidated and very carefully studied over a long period to correspond to the culture and its environment. The equivalent of the over-specialised animal adapted to an unchanging environment. Unfortunately the environment has changed very radically and there is the strong likelihood of the extinction of the species unless there is the recognisable drawing back to leap forward action that is now so obviously needed.

The theories and attitudes of the 'Modern Movement' in architecture become even more clearly misleading if examined as part of a general cultural message system. It can be seen that they do not refer to the new culture developing from the turn of the century but to the one before that, the theories and attitudes do not belong to now at all.

It is quite clear that this 'movement' in architecture never left the Academies in the same way that the painters and sculptors and musicians left theirs; the radical reappraisal never took place and the academy is still a going concern but the question is for how much longer.

The image making approach is now so run down and the students are slowly becoming aware of what is happening elsewhere that there is little to copy in the first formal learning processes apart from the present cult of Victorian Revivalism. The architecturally advanced magazines have almost stopped the traditional publication of architects pristine-just-finished monuments with their noticeable lack of human beings, and are taking a long hard if slightly misguided look at the significance of the changing cultures in the real world outside the profession.

Contrasted to this is the noticeable success of industrial designers and stylists who make efficient constructions and manufactured machinery a joy to handle and possess without confusing their mechanical processes. "Braun" heaters, insides of aeroplanes, or american automobiles can be used, enjoyed, left, or thrown away in a manner that is completely alien to the way that buildings can and are used. The first corresponds to the purpose in our culture and the other does not.

LIFE SYSTEMS

Urban situations or buildings are reflections of life systems; a collection of inter-connected and inter-related phenomena that cannot be separated out into their material components and classified and manipulated according to classic principals without their destruction and death. The whole is greater than the parts. If they are over-organised with sets of figures and statostoca; surveys then there is little room for signification and all appears to be forbidden; like New Town housing layouts or Performing Arts Centers and Civic Halls.

The purpose in the present cultural message is that we know a lot but it is yet to find a fixed structure and so while this process is going on any building or part of the built environment should express this message unless its purpose is very clearly structured, unless its purpose is clearly known and not liable to reconsideration.

17

Why to Design: Control

MACHINES

A machine in the generally accepted sense represents all that is characteristically anti-human. Certain functions in a society might be delegated to certain mechanical devices but any implied resemblances between these and human agents are considered but trivial.

However, in the last few decades there has been a radical change in our conceptions of machines and their possibilities with the growth of communication and control. The idea of the machine has been so enlarged that very few, if any, specifiable characteristics are in principle outside its scope. It has become meaningful to ask not only what function machines may perform for society, but also how far societies themselves can be regarded as machines, and whether the remedies for the breakdown of machines can be applied to the corresponding disorder of society.

Since the notion of a mechanism is never far from that of manipulation, this raises a crop of questions which go to the roots of the traditional conception of social responsibility. As machines could become the means we use to control the environment, not only is it necessary to know something about individuals and societies, as well as something about environment itself. Therefore any discussion must take into account three aspects.

The first is concerned with the nature of the new scientific concepts that have so revolutionised our thinking. The second with

151

some examples of their relevance, both practical and theoretical, to human society and its relation to its environment. Thirdly, some of the curious limitations which beset the world be investigator of manipulator of any society that includes himself.

The possibility of understanding existing social systems and their elements in terms of information flow models does seem to exist and it also suggests the possibility of using the machines as a pre dictive ever-changing evolutionary models. And it is here that it seems a suitable programmed computer could help with the calculations.

There is no doubt that an accurate information flow model of a society could be a powerful tool in the hands of anyone, whether with good intentions or bad. But there is one snag to this idea, and that is it can only give unlimited power if the predictor himself can become sufficiently isolated from the society for which he is doing the predictions.

But however mechanised may be the information system embodied in a social structure, the significant fact that is units are themselves cognitive information systems. This does not mean that the individual human is not explicable in information flow terms as mechanism, but that the behavioural characteristics of units of human society are sensitive to information.

To publish a prediction so strongly affects people's behaviour that it invalidates the prediction. The same considerations expose what Popper has called the fallacy of historicism, namely the idea that human history is inevitable predictable by extrapolation from its past.

There are many human situations where such extrapolation is logically impossible, since the attitude of the predictor himself to his prediction is a part of the data needed to complete the prediction. However scientific the basis of a prediction of the future it is still possible for an individual or a society to take the opposite view and therefore make nonsense of the one taken when the calculation was made.

A complete description of a cognitive information system must include and depend on the information possessed by the units of the system. Any change in information would require a change in the complete description. No complete machine model, or predictive model, of a society is possible, which was equally valid before and after any member of a society heard about it. So here is a basic logical impasse. And here again is found the fallacy of the declared aim of science to take a neutral position and to be unaffected by the attitude that people take to its conclusions and so useless in this situation.

It is obvious that fundamental questions being asked in the social sciences can have no such answers. Objective information cannot be supplied by the social scientist. They are as much a part of the society to which they belong as any other member, and can be no more objective than say a psychiatrist. They cannot set up or offer a special competance to decide that society should pursue one goal or another, but only supplying a certain skill in calculating what may happen if it does.

Here is the growing point in the understanding of social phenomena; what can justifiably be believed by a group must obviously be related to what can be justifiably said by each member of that group. But the logic of the relation between 'we' and 'I' is little understood except possibly in passing by the ethologists. It is one thing to understand the mechanism structure as an information flow system and quite another to understand its behaviour as a cognitive information system.

It is clear that a period of confusion now exists between 'science' and 'value' which applies in our social application of science. Discussions of value not based on authoritarian ideals take place elsewhere but a look at Isaac Asimov's Laws of Robotics might suggest a possible way to go without doing so.

The Law of Robotics as a "value system." 1st: No Robot may harm a human being or inaction allow a human-being to be harmed. 2nd: No Robot may dis-obey orders unless it conflicts with the first law. 3rd: No Robot may endanger or destroy itself unless it conflicts with the 1st and or the 2nd laws.

GENERATORS

The best chance, as things now stand, seems to lie in the idea of creating GENERATORS, which cause deliberate action in order to bring about a new resulting equilibrium, and which consciously sets out to alleviate the most obvious causes of dis-equilibrium. Further problems are bound to arise in consequence as can be dealt in the same experimental, trial and error way.

The object is to try to reach a balance by trial and error experiment that is constantly changing; where there are no once and for all solutions but only the initial stages of a conscious evolutionary process. Such means and ends of environmental change will have to be mentally and experientially prepared by a small nucleus of people in different areas who understand the inter-connection between human behaviour, human experience made manifest, and his environment.

What is necessary is not a collection of specialists working for consensus and compromise solutions; called architects, physicists, biologists, zoologists, ethologists, philosophers, and sociologists, who are only aware of the crisis from their own point of view, but a group who can take the findings of all these disciplines and use them for one central purpose and that is environmental control.

The objective must be to isolate as far as possible the cause of any apparent dis-equilibrium, that is a tendency to disorganisation, be it biological, functional, social, and to plant the 'seed', the generator based on a model of what might happen; to make a conjecture and allow the test of the real world act as the critical refutation. Such a model is most unlikely to last very long, as the construction of any model is bound to be very crude and will change as the generator begins to work and so there will have to be the possibility of monitoring the result as a feed-back. Such models in themselves would be based on the result of experiment in themselves. Any attempt to apply precise solutions to precise problems is never going to succeed. The situation being far too complex in organisation and constant evolution.

As time goes by it is becoming very clear that "natural laws" cannot be found for everything and that behaviour of people and social and cultural attitudes seem to belong to this category. The declared aim of science is to propound conclusions which not yet disproven at least for a moment, regardless of the attitude people take to them. It now seems clear that many questions now being asked in the socio-environmental field have no such answers. A fact that remains very much unpublished particularly in the social sciences. Even the natural sciences. Even the natural sciences are finding it extremely difficult to find definitions that can have models or analogies constructed for them and now prefer to make definitions in the order of "matter cannot be known except mathematically" and "an atom can only be described in terms of its behavior not as a rigid structure but as a series of energy states and resonances." It is a pity that social scientists cannot take a leaf out of the same book.

Environmental modification must aim at the intensification of life, both by strengthening its roots through better functional arrangements that raise the man-environmental relationship to the level of genuine psychic experience providing the stimulation and identify to both individuals and social groups.

No future world-city or Utopia can usefully be presented; more often than not the image of an ideal progressive future is only attempt to escape from the most decisive problems of the present. Any useful work must be a continuous process originating in the matrix of present conditions whatever they are, in the highly developed West or the underdeveloped rest of the world. The environment must assume an unforseeable multitude of functions and forms. To deliberately effect such an evolution rather than waiting for it to happen would mean that instead of resigning ourselves to the tyranny of historic determinism, we can impose a continuity if space and time.

ENVIRONMENTAL DESIGN MUST BECOME FUTURES RESEARCH.

18

What to Design: Information Systems

SYSTEMS

Although the exact numbers of fundamental particles are somewhat in dispute, it is supposed to be in the order of 10^{79}. This is incomparably larger than any forseeable human or animal population. Yet nature controls the stability of processes among fundamental particles by a very simple device; it does not attempt to manipulate them all at once. It builds them up into stable sub-units and then assembles these sub-units into more elaborate units and so on. Nature builds up complex structures by working with hierarchies. There is a standardisation of sub assemblies with built in rules that permit a great amount of variation but only in a limited number of directions and on a limited number of themes.

It seems that it is impossible to find any natural phenomena which proceeds in any other way; large stable structures can only be built from small stable structures. This is just as true of clusters of stars as it is true of families and communities. Wherever there is a human community it has to start by being built from small self-coherent groups even if it adds up finally to a group of 3-400 breeding numbers which can be called the basic village, even within a city. The integration of these human groups, step by step into national and international groups are very difficult, but there is no doubt that this remains the only way of creating large human societies that are stable. Thus any structure of such societies, for instance, that are built in the future will be achieved not by uniformity but by integration of varied and individual groups in a hierarchic form; an organisation system 'governed' by 'information-flow system' or 'purpose-generating system.' The more uniform and flat the system the more vulnerable to collapse.

ASSOCIATIVE SYSTEMS

In formal organisation and in practice constructed systems are usually designed as 'associative systems'; that is to say that they are made up of: PURPOSE GENERATING SYSTEMS. Purpose generating systems concern the flow of information, messages, about content, purpose, intention, the reason "WHY" the construction exists, coded messages about the cultural behaviour. ORGANISATIONAL SYSTEMS. Organisational systems concern the organisation, language, that conveys the message in terms of 'cultural space' and the arrangements that reflect the common usage of those "spaces."

An associative system is considered to be a system because of the overall purposefulness of the organisation. The deletion of any part of the organisation system is likely to diminish the purposefulness by cutting down the total possible message; too much organisation system will lead to cliché.

Every kind of built environment can be considered as inter-acting with the people who inhabit and use it. It both dictates the way it should be used and is there in response to the use that originally purposed its construction. These uses are always changing in response to each other. The constructions last over a longer time sequence than their original purpose, so the first can get out of sequence or step with the other and can go out of equilibrium.

These constructions seem to fall into two main types of a combination of the two; cultural information systems that are intended to ensure a particular type of behaviour as laid down by the culture, as in churches, parliament, or other 'formal' situations; and those that are intended to respond to changing behaviour and norms or 'informal' situations; all three possibilities expressed by the spatial arrangements internally, corresponding to the signification and structure of the message. These observable patterns of interaction can be conceived as a system of information that can be analysed and described in existing historic and existing situations.

"WHY" SYSTEMS, "WHAT" SYSTEMS, "HOW" SYSTEMS

These two systems describe 'why' the building is there and 'what' it is for; included in these two systems are the cultural messages, including all matters of purpose behaviour and aesthetics. Completely separate from these systems are the mechanistic systems that are intended to provide the controlled climate and envelope for the cultural message systems; the mechanistic systems are 'how' it is to be down and they themselves have organisation systems but their purpose-generating system is laid down by the 'why' and 'what' descriptions.

GOAL-DIRECTED

Therefore an architectural work can be seen as a collection of information so ordered and unified to perform in a certain manner; it is goal oriented, it is there through deliberate choice and it is a system because the information is managed in a hierarchic form and its mechanistic systems correspond to the same hierarchies or systems.

ARCHITECTURE DEFINED

Architecture consists of an organisation of deliberate and carefully selected items of cultural information; and so therefore architectural works and the built environment which should be considered as synonomous, and their methodology of 'creation' can be studied as constructed systems emboding cultural and social information.

Information is stored in various forms and transmitted. Architecture can be considered as a language of human beings communicat-

159

ing with and being communicated to by nature. Spoken language is the most articulate but not the only language there being many non-verbal message systems through physical behaviour which is a combination of message systems indicating different needs and satis- factions.

ARCHITECTURE AS A LANGUAGE

Architecture can be thought of as a proto-language or as a funda- mental form of expression and communication occurring outside of or before 'words.' Significance in proto-languages is usually expressed with considerable redundancy, and explication is only possible if the surrounding circumstances are considered. Meaning can often be less succinctly expressed in words, but the effect on human behaviour is still very storng. There is always a strong reaction to spaces, shapes, colour and touch which if less conscious as words is more direct and spontaneous. Thus architecture is a form of cultural message which can and does act as a controlling influence on society. Architecture is communication and communication culture.

INFORMATION SYSTEMS

The central notion of the science of communication is that of information flow. Information is said to flow from A to B when the spatio-temporal form of something happening at A determines the form of another at B without necessarily supplying the energy for it. Thus it can be said that when a man pushes a door bell he sends information to the kitchen, because the form of action has de- termined the form (the timing) of the ringing of the bell, regardless of the energy source.

By analysing the logical process by which a form is determined, it has proved possible to define useful numerical measures of in- formation flow. But for present purposes it is sufficient to recognise the qualitative distinction between explanations in these terms and those in terms of traditional physics. It can be said that whereas

160

physics looks for explanations in terms of dependance of 'form upon form' or 'system upon system.'

Systems whose functioning depends on information-flow are called 'information systems.' By abstracting the information-flow map of such a system important common features can often be discerned in extremely diverse situations; as well as leading to an understanding of their working. Instead of loose appeals to analogy it becomes possible to apply common principles to the understanding and regulation of different situations as well as making it possible to measure the amounts of information available.

SYSTEMS

There are two ideas hidden in the word system; the idea of a system as an organising principle and the idea of a message and the idea of a purpose-generating system.

ORGANISING SYSTEMS

An organising system is not an object but a way of looking at an object or series of objects. But as objects are unobservable and only their relationships among objects are observable, all that can be considered is the hierarchical structure and interaction among the parts from a particular pre-defiened point of view. There can be no absolute point of view.

PURPOSE GENERATING SYSTEMS

A purpose generating system is the part of an information system that conveys the way the parts should be combined to achieve a specific purpose.

Every organisation system is governed by purposeful goal directed information to make new associative organisation systems; new purposes have to be invented; to change conversely the purpose of an existing system can often be discovered by the analysis from existing associative systems.

The analysis or construction of associative systems has nothing to do with the way the associative system message is transmitted. The medium has nothing to do with the message. The techniques used to construct the models of associative systems are irrelevant to the construction of the mechanical techniques for transmitting the system as a message.

Systems deal with stability, which is a property of hierarchies.

The formation of systems is not by the addition of parts but something of an entirely different order. In an aggregation it is significant that the parts are added together; in a system it is significant that the parts are arranged so the whole is greater than the sum of the parts.

A system is an abstraction, it is not a thing but a special way of looking at things. Even though something is called a system this does not mean that it is ever viewed as a whole in its entirety. The area of a system to be viewed is abstracted from an even larger and apparently endless system. A system is not synonomous with an object but an abstraction that has to be deliberately identified, and cannot be so found by the process of induction, but must be defined before the event; not after when it would prove impossible. The systems most important characteristic is to point out the inter-action of all things and the impossibility of classifying them into sets of objects.

In architectural terms, purpose generating systems convey the content of a 'construction' and the way the content is to be structured; and answers questions as to 'why' the construction is there; the cultural need for it; and from the structure of this message the answer to the question about 'what' is to be there. This then gives guidance to 'what' is constructed.

ANALYSIS

Analysis of an existing construction can then be seen as the reverse of this procedure. The analysis is concerned with the identification of the associative system embodied in the construction, or the various associative systems that may have been built up over the historic time of the construction, as far as possible so that the cultural human behaviour can be deduced from it. This is obviously much easier if possible at all from buildings that have not been altered in any significant way since they were built. The amount of information is dependent on the amount of structure or signification in these messages through the associative systems. The appreciation of these messages, their signification and structure, and the emotional response to them is what is understood as beauty or aesthetics.

STYLE, STYLING

On the other hand the way the 'construction' looks, which can be called styling or 'style,' depends on cultural fashion; that is the way people like things to look as an amalgum of old accepted forms and novel new acceptable associations, as in clothes, cars, advertising, packaging and so on. The intention is to personal identity as a forward looking person, but not so much so that they can be classified as excentric; so a general consensus is necessary culturally but will have nothing to do with literary or academic values of cultural historicism. Essentially fashion or styling is of now and is constantly changing, as visual stimulation value and novelty is what is important.

AESTHETICS

In buildings, which are especially caught up in historic cultural values, the effect of fashion is much less noticeable in exteriors or the total fabric, but is much more predominant in interior decor for shops, restaurants, cinemas, and other areas of actual human experience concerned with buying and selling and generally living. In this way very old buildings have often changed their usage and interiors while retaining the main fabric unchanged for hundreds of years. The need to tear down buildings very often arises because it has been argued that it is less expensive than adapting the existing to a new use, or another building could have proved more suitable; but very often the demolishing takes place because architectural theory and advice demands that it is not possible to be modern without total reconstruction in a new idiom, being unable theoretically to separate the construction of the fabric, the 'styling,' and the message, its structure and signification; its 'aesthetics.' If this difference between 'styling' and 'aesthetics' were more thoroughly appreciated there would be far less destruction of historic towns, slowly adapting buildings rather than total destruction. Evolution not revolution.

EVOLUTION, MUTATION

Nature works by slow evolution, slowly adapting the existing to fit slowly changing environments. A revolutionary and successful mutation is very rare and the same thing happens, or should happen with buildings, unless the general cultural-environment becomes quite different. In these circumstances the mutation does not come about in the superficial styling, but in the purpose generating system; the sub-assemblies, which are seen in animals as arms, or wings, or fins, do not alter in response to a required change in their styling; but in response to a change in *purpose*. New buildings can retain their subassemblies, in the same way while changing their purpose in response to cultural change of purpose. This is the meaning of "ORIGINALITY" in architecture. It involves in redefinition of the purpose, not an anarchic confusion and battle between styling; that is current fashion in architecture as a poor sort of historicism and technological imagery; and new purpose. Trying to change purpose with arbitrary architectural styling, which goes a long way to explaining the very apparent chaos on the contemporary architectural and town planning scene. There is an implicit and explicit fundamental belief in these professions that a change of 'styling' has some equivalence to changing cultural purpose. A belief that the acquisition of a few beads and some long hair makes a hippy in mind and

164

behaviour, or plastic surgery will make an unpleasant woman suddenly pleasant.

EQUILIBRIUM

Equilibrium is another way of saying structure or organisation.

The object of any design endeavour is to see that there is an 'imbalance in the man-built environmental system and try to restore this balance; to separate out those things that are pulling too hard in one direction so that they distort the balance the organisation; like building motorways without considering where they go and what to do with the cars at either end. Many areas have no great imbalance or lack of organisation, and these should be left alone; other areas need full-time attention, there are no panaceas, each problem has to be considered on its own.

CONSTRAINTS

What should happen is that any small bit of the area is first studied as a little bit which each person knows something about and then this is extended with new information until the whole area is understood and the unbalanced disorganised part understood in the terms of that which surrounds it. If a triangular diagram is imagined to describe the whole system in balance and then one or two bits taken out and changed or expanded without taking the rest into account, then the whole thing will be distorted. The object is to build up as much knowledge as is possible about all the context and see what generator is needed to restore the balance; a building, or a political decision, a painted sign. Each could 'solve' the problem. What must be aimed at is a restoration between people and their things.

CONTROL STATEMENTS

The discovery of the constraints of the areas under observation is in effect a CONTROL STATEMENT of the problem. It deliniates how far back up the systems it is possible to go and still be able to effect the purpose generating system; how much can be decided in the 'why' question area, what the content/signification is going to be. A control statement defines the organizing principle the purpose and the nature of the information to act as a generator.

$$19$$

How to Design: Procedure

The designer is essentially a solver of problems, as presented to him by Nature, or the task, or the materials he is using, or the people who will be using his end product.

The model of the man which emerges when considering a man as a problem solver, is associated with a date processing, and he can be seen as a hierarchy of systems in which he receives data, processes data, and puts out data.

Problems cannot be solved, or defined, or be seen to be problems without information. The more complex the problem the more complex the information that gave rise to it and vice-versa; the more complex the information available the more complex the problem. Problems only seem simple because not enough information has been collected. As the world information store increases so does its problems and their solutions.

Problems are what we have to make decisions about, that is, there can be detected a lack of organization or an imbalance in any system that is being looked at. If things are left in a state of imbalance they will not right themselves automatically but tend towards disintegration and self-destruction. This is why it is always necessary to reverse the direction towards entropy if only locally, by introducing new information.

Thus it appears that through the application of information theory, and feedback design machines can be created to respond to their environment not unlike living organisms, even if we still have to

167

call them buildings. Design strategies can be worked out to bring this about.

DESIGN SITUATIONS: EMERGENT, ESTABLISHED

If a continum of situations in which action can occur it will be found that at one end can be classed as 'established' and the other end as 'emergent.' An established situation is one in which an action relevant environmental condition is specifiable and predictable and all action relevant states are likewise. All available research, technology or records are adequate to provide statements about probable consequences of alternative actions.

In contrast an emergent situation is one in which some of these conditions do not prevail. There is a connection between these two kinds of situations and the notion of knowledge by experience and by learning. Situations met with in everyday life are emergent; those through learning, established.

Designers of utopias are concerned with established situations; same with social surveys of a population, or a controlled experiment, or work by a skilled or unskilled worker.

Painting a master-piece involves dealing with an emergent situation as does all creative art and problem solving, including playing chess and designing for a complex environmental situation.

In a design procedure it is important to note that some system may be required to deal only with established situations while others will be required to deal with only established emergent situations; some with a bit of each.

Questions are often concerned not only with what problem is being solved or why it is being solved but what kind of solution would be accepted as satisfactory. Differences in approach involve implicit if not explicit differences in the answers. They imply gross differences in methodology and technique. Bososiau identifies five different basic methodologies as follows:

FORMALIST DESIGN APPROACH

This is characterised by the implicit and explicit use of two types of models: the 'replica' and 'symbolic.'

Replica models, 'image-making' or 'master-work,' provides pictorial representations that are material or tangible and seem to work like the real thing with a few concessions to change of use and place.

This is basically the 'functionalism' of the 'modern movement' in architecture. Included is the superficial imitation of the way ships and machinery looked and more recently the visual appearance of rockets, space flight equipment, electronic technology, sound and light amplification and 'plug-in' component systems.

In the face of the complexities and traditional disgust of all things mechanical or electrical, another group has taken an alternative by rejecting all these influences and indulging in an historic revivalism, neo-Classical or neo-Victorian deliberately choosing a building technology little different from that used in Babylon; and another group in the idea of trying to express the building's purpose by the way it looks; an aircraft terminal symbolised by buildings representing flight; the 'expressionist' approach; a further group is forever finding ways of trying to indicate that 'truth or beauty' are manifest, but only to them and other members of the fine Arts point of view.

Symbolic models, the mathematics-are-difficult-and-must-be-the answer, are intangible like ideas and concepts and abstract symbols to represent objects and manipulate them; none of which in any way resemble the appearance or necessarily the behaviour of the objects under discussion in the real world. They use lines and arrows to symbolize information flow, or statistical survey material, statistical and other mathematical models, operations research, various managerial techniques and systems analysis.

This is the current fashionable design strategy, being much used to develop new 'Design Methods'; as if these models represented real things and not a predetermined set of figures to be manipulated for a desired end; the discussion of which is not taking place at the same time.

Both approaches are firmly rooted in the mechanical and pre-mechanical age and their theories and methodologies. Functionalism is obsessed with the wires in their boxes and where they go in preference to the electricity that flows through them and what it is doing; a failure to understand that the wires are there for a purpose and if solidstate circuits will do the job more efficiently then the wires will go whatever they look like. Its protagonists fail to appreciate that this kind of approach to technology is far from liberating, that this kind of formalism must of necessity ignore the people inhabiting them, that the medium does not represent the message and that technology is the most changeable component, not human or cultural behaviour. Furthermore neither method can make any sense except in highly probable and predictable situations; they are based on historic precedence and can have no predictive value; they are sterile.

HEURISTIC DESIGN APPROACH

This is characterised by the use of principles as guides to action. It is not bound by pre-conceptions about situations which any investigations might uncover. Its principles will provide action guides even in the face of completely unanticipated situations, and in situations that occur where there is no formal model available.

These principles are of necessity usually global and vague, both to suit the situations that arise and the various different types of people who want to use them, insisting on such things as an ideal society and all must accede with these concepts or be considered out the pale and in some way inferior.

This is the Beaux-Arts design approach, and that of the architect as social and moral reformer; and the heuristic design methods that once found can be applied indefinitely to all problems.

OPERATING UNIT DESIGN APPROACH

This is characterised by the rejection of models or selected principles and goes for a careful selection of people or machines which
possess certain performance characteristics, so that whatever emerges
will incorporate these previous desired solutions that these units
provide, and are also governed by the limitations of the units flexibility. The extent of the pre-determinations and performance specifications will govern the freedom of operation within the system.
It will also allow the original programmer to be in full control as
he can determine what goes into the system. It is like joining a professional institute, you will always know where you are even if you
cannot in exchange do what you like.

This can also be called the 'catalogue' of parts' approach. The
final outcome is that a selection can be made of all the rules and
regulations that might apply to a building; legal, financial, ergonomic,
structural, industrial, and so on; put into the store of a computer
and with the aid of various visual displays allow the architect free
reigh for his 'creative' games; great 'aesthetic' evaluations and in-
credible flow of intellectual abilities, unfettered by the sordid side of
being an architect that now takes up 80% of his present time.
Coupled to this particular utopia are two further ideas, one concerned
with the 'cyborg' building that responds to its owners every whim
and desire, and the attitude to industrialised buildings as a perfect
set of parts designed by architects that once readied for mass-pro-
duction should never be changed and so involve the architect in extra
work away from his creative computer, in fact the very way that they
are to be designed and made would not allow them to be changed
anymore than a brick.

Experiences in this way of things are not to be opened up but
are to be restricted to a statistical norm based on questions that concern the traditional ideas and practices and methods of design, as
if there was nothing wrong with them. The computer is to be developed and financed to a degree that would make them only availa-

171

ble to a few wealthy offices and the building industry is to be geared to this way of working without so much as asking their opinion.

The theoretical use of the computer totally misunderstands the nature of information and the creative process, assuming an analogy equivalent to buying a car that it is assumed will take you wherever you want to go without knowning how to drive or where you want to go. Without knowing what is to be done, the computer is only a machine capable of very rapid digital computations of high accuracy, which have to be converted from analogs unless of course all analogs are aleardy pre-determined in which case it all becomes a giant random search or lottery among known components. Computers have very low flexibility where human minds have low accuracy but high flexibility; they complement each other, one determining what shall be done and the other working it out.

PRAGMATIC DESIGN APPROACH

The characteristics of this approach is that it resorts to no models, principals or operating units, preferring to proceed from the view that present reality is the only constant in the equation. It begins with the review of everything that is available and happening. The next stage is to make some sort of move and then modify according to the consequences much after the pattern of evolution and information feed back except that the line of least resistance is always taken. It constitutes a search for an easy solution and there is no clear idea or need for an idea of the future.

This approach is most closely associated with those who out of ignorance, stupidly or lazily prefer expediency to deliberate action; 'muddling through,' 'something will turn up,' 'it's fated anyway and will sort itself out,' 'I am only doing it for money or fame.' No particular example for this deging strategy is needed as about 80% of all the built environment is achieved this way. There is a complete lack of predictability and relies on temporary solutions, fixed repertoires and static environmental conditions.

EXISTENTIALIST DESIGN APPROACH

The characteristic system unlike the others is to take the human situation and the reality of human existence and to derive meaning from all human activities and behaviour. Categories and principles like time and space are insufficient for understanding this fact or existence. Computers and their schematic and abstract solutions that follow in their wake are instruments of retrospection not instruments of progress. Individual existence demands change, incompleteness and lack of closure that the first four methods and approaches cannot provide and even deny. The question is how can these human experiences be incorporated into a new logic of architecture and the built environment clearly reflect them.

'AND,' 'OR' RELATIONSHIPS

The designer's task is made much easier if the mass of information coming in can be reduced to make the task simpler and therefore make easier to understand. His objective is to reduce the number of choice of action to one, either this or that. This is the explanation for the desire to pre-conceive answers to problems because it immediately simplifies the alternatives by ruling out many possibilities. Therefore the designer can be seen as searching for ways of removing possibilities until there is only one left at the end. This can be reduced to the concept that everyone is trying to reduce all their 'or' relationships to 'and' relationships; reducing external analog inputs into digital 'yes' and 'no' for mental processing and the fitting of expectations and experience in the manner of everyday living; a designer bases his methods, having never explored their nature, on the same pattern and procedure.

'And' relationships are called 'conjunctive'; 'or' relationships are called 'disjunctive.' But conjunctive relationships reduce information content by increasing structure at the expense of signification and vice-versa.

173

Useful architectural problems are not conjunctive ones needing single variable answers but disjunctive ones needing multi-variable answers. 'Or' has many more possibilities than 'and.' When some designers and designs are said to be better than others it is often because they work in the disjunctive area; the poorer in the conjunctive area. To design well is to avoid converging down to one single solution but to state the whole range of possibilities and solutions available and then select the most appropriate one from them or leave all the options open.

THE LESS STRUCTURE THE MORE INFORMATION: THE GREATER THE CERTAINTY, THE MORE THE MES-SAGE MEANS.

What needs to happen is to observe in the structure of each problem the juxtaposition of relationships so that the designer can perceive disjunctive relationships easily between aspects of each problem, so that the whole method of presentation to himself must not be in such a form that he cannot see that it is there. The structure of a problem presentation technique is easily used by a bad designer to structure the problem itself. He is so willing not to have any structure in, because it reduces the possibilities; if you cannot draw it, for instance, it is said not to be architecture or an architectural problem. Drawing on paper rules out the possibility of thinking and working in three-dimensions; the T-square rules out many forms outside the use of the right-angle and enshrines the plan, section and elevation as the only way of acceptable problem solving and presentation in architectural problems.

BUT THE MORE STRUCTURE, THE LESS INFORMA-TION: THE GREATER THE UNCERTAINTY, THE LESS THE MESSAGE MEANS.

174

STRUCTURING PROBLEMS WITH DESIGN STRATEGIES

Techniques of design have to be assessed on the basis of how well they do not structure the problem; they must be techniques that do not structure, the analog with a technology has to be removed from the technology, and the analog with something else introduced. To take the drawing board out of architecture would cause the whole organisation and working method to collapse; to stop drawing buildings would have the same effect. Architecture is structured by these techniques at present and there is no possibility that computers, or new design methods can solve this problem; there is so much traditionally based structure that there is no room left for significa- tion just at a time when the culture is demanding greater amounts of it.

This is a way of looking at design objectively with predictive pos- sibilities is a particular design strategy or technique going to help or not, is it going to structure the problem? It is not a matter of whether it produces solutions or not of some sort, which it is bound to do but does it move away from this subjective position and predictively help?

GODEL'S THEORUM AGAIN

SEMANTIC PROBLEMS CANNOT BE SOLVED IN ANA- LOG TERMS WITHOUT EFFECTING THE RESULT, BUT ONLY BY ADOPTING SOME META-LANGUAGE WHICH FOR THE TIME BEING IS BEYOND QUESTION.

Mathematically speaking information is as fundamental as energy; information and energy can be causally related. They are both abstractions measured as input and output, not as an object. As voltage is the potential difference across a wire, so the measure- ment of uncertainty before and after an event is call information;

although no one knows exactly what it is. Uncertainty is analogous with entropy, information decreases entropy by increasing organisation and structure, and information can be measured as a quality. Structure and quality can be thought of like concepts of melody or beauty and can in consequence be measured. So what were originally thought to be qualitative and can be discussed as long as the qualities are known.

AESTHETICS CAN BE MEASURED IN TERMS OF REDUNDANCY AND SEQUENTIAL AND MULTI-VARIABLE UNCERTAINTY.

The more redundancy the more structure and therefore the more easy to understand. To reduce everything to 'and' is to loose information and allow disintegration. In any given social situation there is a lot of information from a lot of people and needed by a lot of people and therefore it is only describable in disjunctive terms, in spite of many recent efforts to reduce it to conjunctive statements. So any constructed envelope must not restrict the disjunctive possibilities and these creations must be at the highest level of capability in order to respond to disjunctive demands.

The more structure/redundancy in something, the less uncertainty of information transmitted and the easier to understand. A nursery rhyme is easier to understand than Shakespeare because of greater redundancy, as the work of Mozart when contrasted to Wagner's shows. Norman architecture is so good for its purpose because it is so boring that there is nothing left to do but listen to the preacher, or create internal day-dreams and fantasy, as time cannot be left unstructured. Baroque churches have the opposite effect. Norman is so redunant that there is no uncertainty left; as it is so structured. Baroque churches have large masses of uncertainty. Some buildings need large information content, some do not. Clever students in schools of architecture have realised this by trial and error and will always choose town halls or concert halls or ballet schools or whole town centers because they can get a great deal of information into them and therefore show their teacher just how much they have learned. The less clever choose churches or factories because their

176

information is not in the building construction and its mechanical services.

AESTHETICS

Aesthetics is concerned with the appreciation of the structure and organisation of messages from the environment or from other people about the environment and co-incided with the organisation of each individual brain. Some get great satisfaction from the low information but elaborate structure of the music of Bach, others from the rambling disorganisation of Wagner. These simulations give rise to pleasurable emotion which are called aesthetic experiences and they have nothing to do with the materials and structure of the media that conveys the messages. This is the complete invalidation of the Modern Movement in architecture. The pleasure of a painting is not in the same way the paint is made or in a sunset in the fact that the sun is burning hydrogen.

AESTHETICS ARE NOT MEASUREABLE IN TERMS OF DIMENSIONS AND PROPORTIONS BUT IN INFORMATION AND REDUNDANCY.

The normal procedure in designing can be seen as trying to solve the problem by trying to push everything into the Motzartian idiom, of high structure and low signification, rather than the other way around. The question is really about keeping all options open, exploring all the possibilities and alternatives, and then choosing if necessary; not a question of going for the first possible opportunity that presents itself to close down the options. The whole thing has to be opened up to find the underlying structure or pattern, but not one that reduces the information to nil, to cliché, but one that allows anything to take place.

SIMPLICITY LIES ON THE OTHER SIDE OF COMPLEXITY. THE SOLUTION ON THE OTHER SIDE OF CHAOS.

So here is presented a means to measure the difficulty of a design problem mathematically, something that numbers can be put to; the degree of uncertainty of possible solutions to a problem; and here is the meta-language that is also needed. The designer is trying to find the structure, or redundancy. The measure of the difficulty of a problem is the measure of the maximum number of solutions, minus the number of feasible solutions, which are the result of taking everything into account equally. This is the definition of how much less and it is directly proportional to the amount of structure between these boundaries.

DEFINITION

DESIGN: IS A PURPOSE DIRECTED CONTINUOUS PROBLEM SOLVING PROCESS AS A SEARCH FOR CORRELATION BETWEEN MULTI-VARIATE UNCERTAINTY BY CONJECTURE AND CRITICAL REFUTATION.

Good designers intuitively produce high information content solutions. They can be looked at time after time and still give satisfaction; it does not seem boring because more and more can be discovered in it; but no more can be got out than was originally put in; that is no more than the original designer was capable of putting in; inanities are very redundant containing little if any information. A design solution that contains small pieces of banal information about situations that everybody is familiar with and are already bored by has to be contrasted with a solution that contains new and unexpected situations, in depth. People seek information for their own identity, security and stimulation. Such a designer must be acquainted with what is going on in their own culture in the present and be aware of all its knowledge.

The task is very complex because it is necessary to have some parts of the built environment with high content and some with nil content because any information that the building contains, for instance, will have an effect on what happens within it. It is of paramount importance to identify this quantity of needed information, and

178

this forms the first part of the initial design strategy and search for a correct match; in teaching buildings or factories for example, the information content should be very low because cultural information in this area is very dynamic, flexible, and constantly changing; it is very unstructured.

OBJECTIVES OF DESIGN

In this way entropy measures; uncertainty measures of what is going to happen; what is most likely to happen; not all peoples space needs but what action will effect people the LEAST, coming into a given area; it is a negative thing like all science laws, rules of behaviour that cannot take place; not must take place. Space must not be more structured, there is no need to find out what everybody wants, as if this were a possibility, but structured the least. But there must be enough structure to ensure that the activity is clear if the construction is for a specific purpose. The more structured the activities the more certain they are, the more redundant they are, the more predictable, the more specified, the more goal-oriented and the more carefully they can be organized.

The more carefully design can be done the more carefully it is done. So what initial strategies are defining are how far a design must be structured; living room entrance halls need less structure than kitchens. The total of uncertainty is greater; there is less definable activity going on, and a lot more different things could go on than in a kitchen. The attitude is away from the most closely designed to the least closely designed; the least that can be gotten away with in the cultural context or the ergonomic context.

SUMMARY OF WHOLE ARGUMENT

ARCHITECTURE IS COMMUNICATION BETWEEN MAN AND NATURE

COMMUNICATION IS THE TRANSFER OF INFORMATION TO ELICIT RESPONSES AND CAN BE TREATED AS ENERGY.

THIS ENERGY FLOW IS BETWEEN MAN AND MAN: MAN AND THE ENVIRONMENT: ENVIRONMENT AND MAN.

MAN IS PART OF THE ENVIRONMENT. IT IS MAN'S EXPERIENCE OF IT AS A FLOW OF ENERGY STATES, PATTERNS WHICH ARE CALLED INFORMATION AND WHICH TEND TO REVERSE ENTROPY TEMPORALLY.

MAN IS A LIVING SYSTEM THAT SURVIVES ON THE INTAKE OF NEW INFORMATION/ENERGY FROM THE ENVIRONMENT, HE CANNOT DO WITHOUT IT.

IF THIS ENERGY FLOW IS ACCELLERATED OR CUT BACK IT WILL IMMEDIATELY AFFECT THE ORGANISM, LEADING TO CHANGE OR STAGNATION.

INFORMATION IS CONVEYED BY MESSAGES WHICH HAVE TWO DIFFERENT PARTS: THEIR CONTENT; WHAT THEY SIGNIFY, THEIR STRUCTURE: THE WAY THEY ARE ORGANISED, THEIR LANGUAGE.

The more recognisable the structure, that is to say, the simpler the language, the easier it is to see what it signifies, to understand it, but the less information it contains; The less entropy it reverses.

THE AMOUNT OF INFORMATION POSSIBLE IN A MESSAGE IS MECHANICALLY LIMITED, SO TOO MUCH STRUCTURE MEANS TOO LITTLE CONTENT.

MESSAGES ARE STRUCTURED IN SYSTEMS, THEIR CONTENT COMING FROM PREVAILING CULTURAL ATTITUDES OF THE SENDER OF THE MESSAGE.

Purpose-generating systems describe content/signification. Organisational systems describe structure/language.

The first contains the rules about the way the parts should combine or recombine. The second is about the way the parts are combined.

ARCHITECTURE SHOULD BE THE STUDY OF THESE ATTITUDES AND THE WORKINGS OF SYSTEMS SO AS TO BE ABLE TO FORMULATE APPROPRIATE CONTRUCTIONS FOR ACCOMMODATING THEM IN A CONTROLLED CLIMATE.

ARCHITECTURE IS THE PROVISION OF A SUITABLE TOTAL ENVIRONMENT FOR THE EXCHANGE OF CONSTANTLY EVOLVING CULTURAL INFORMATION.

Within any system whether purpose-generating, organisational, or mechanical-constructional, there are three functions or roles; primary/purpose, monitory/origanisational, service/mechanical.

The way a system works does not have to be known; only what is put in and what comes out. The objective is to modify the input so that a required output can be achieved. This can be simulated in a model or real life situation.

Example of the way the argument works out:—(to be read in conjunction with the diagram over)

MODEL OF A SYSTEM DESCRIBING AN EXISTING SCHOOL. TAKEN FROM A WHOLE SYSTEM OF SCHOOLS, WHICH IN ITSELF RESPONDS TO AN EDUCATIONAL SYSTEM THAT CANNOT BE QUESTIONED, WHICH IN THIS CASE FORMS THE META-LANGUAGE.

INFORMATION SYSTEMS DESCRIBING THE CONTENT, THAT IS THE WAY THE SCHOOL IS VERBALLY SET OUT AS AN EDUCATIONAL IDEA, AND SECONDLY THE WAY THAT IT IS ADMINISTERED, WHICH DESCRIBES THE EXACT NUMBER OF ROOMS, ETC.

BELOW THE LINE OF INTERFACE THERE ARE THE MECHANICAL SYSTEMS THAT CORRESPOND TO THE INFORMATION SYSTEMS ABOVE. THEY CAN THEREFORE BE SEEN AS THE CONSTRUCTION IN THE REAL WORLD OF THE INFORMATION SYSTEMS.

BELOW AND ABOVE THE LINE ARE EXACT MIRROR IMAGES OF EACH OTHER. THE IDEA IN THE HEAD DESCRIBES THE BUILT SCHOOL AND THE ANALYSIS OF THE BUILT SCHOOL WILL DESCRIBE THE IDEA IN THE HEAD. THE CIRCLE IS CLOSED. THE SOLUTION IS TO THE PROBLEM SET. TO ACHIEVE ANY ORIGINALITY THE SETTING OF THE PROBLEM MUST BE CHALLENGED NOT THE SOLUTION.

TO ACHIEVE A BETTER SOLUTION THE META-LANGUAGE MUST BE CHANGED NOT THE PROBLEM/SOLUTION, BECAUSE THE ONLY ROOM FOR ACTION HERE IS TO INTERFERE EITHER WITH THE CONTENT/ADMINISTRATION SYSTEM OR THE MECHANICAL/CONSTRUCTIONAL SYSTEMS, BOTH OF WHICH SHOULD BE MADE AS EFFICIENT AS POSSIBLE IN THEIR OWN TERMS AND NOT DISTORTED BY OUTSIDE CONSIDERATIONS.

A GENERAL PROCEDURE FOR APPROACHING ARCHITECTURAL PROBLEMS.

IDENTIFICATION OF OBJECTIVES: (Answering "Why" questions) The first objective is to explore a larger total system in order to discover why the input and the output are not matching in the way that is desired. This may be a constructional problem or it may exist elsewhere in the total system.

The search is for the meta-language which will define the content, the signification, the idea, so that a Purpose-generating system can be constructed.

THE META-LANGUAGE IS THE IDEA OF EDUCATION

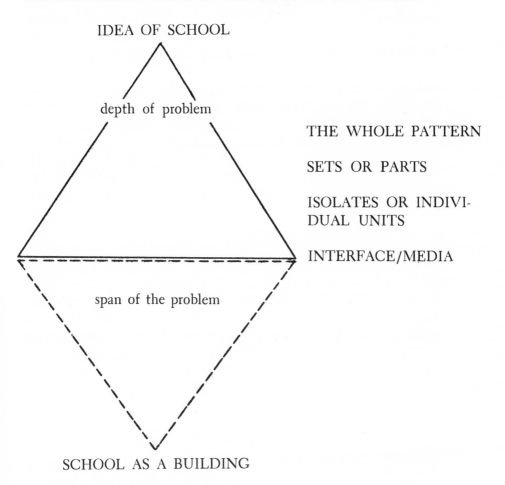

IDEA OF SCHOOL

depth of problem

THE WHOLE PATTERN

SETS OR PARTS

ISOLATES OR INDIVI-
DUAL UNITS

INTERFACE/MEDIA

span of the problem

SCHOOL AS A BUILDING

This system is call the Control Statement and not only defines the problem but also the subsequent strategies for its resolution.

SEPARATION OF FUNCTIONS: (Answering "What" questions)
The second objective is to allocate the policy administration/monitoring system; that is the spaces needed and to describe the performance specifications for the surrounding construction and climate control. The search is for approriate structure for the message, so

183

that the structure/organisation system can be constructed. This is the area that Architects call Analysis/Synthesis. But this area should be controlled only by considerations of greatest efficiency of parts to the whole and should include all outside constraints of laws and finance which should be written into the performance specification.

ALLOCATION OF FUNCTIONS: (Answering "How" questions) The third objective is the physical construction of the message. It is purely mechanical and only concerned with building the required information systems, in response to Performance Specifications laid down.

WHY QUESTIONS
The source of these objectives describes the apex of the system and occurs where purpose generating decisions are outside the jurisdiction of the designer. They are concerned with NOTIONAL metal constructs and RESPONSIVE systems.

WHAT QUESTIONS
Concerned with structuring the message, appropriating the correct amounts of structure/signification or aesthetics. They are concerned with CONCEPTUAL mental constructs and PERFORMANCE systems.

HOW QUESTIONS
Concerned with the appropriate choice of structural and mechanical building systems for the greatest efficiency without extraneous considerations whatsoever. They are concerned with PHYSICAL constructs.

The notional and controlling idea for selecting and organising performance systems

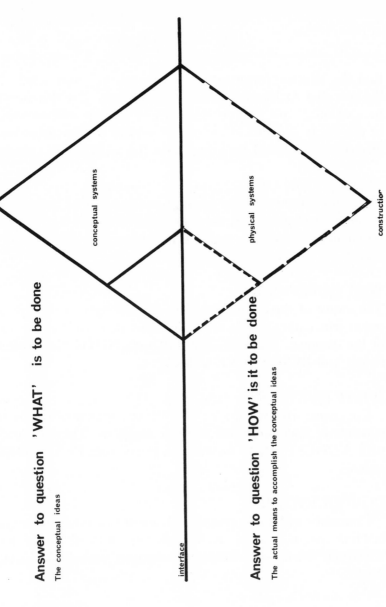

PERFORMANCE SYSTEMS.

1. administrating/monitoring.

2. servicing.

Answer to question 'WHAT' is to be done

The conceptual ideas

idea

conceptual systems

interface

physical systems

construction

Answer to question 'HOW' is it to be done

The actual means to accomplish the conceptual ideas

PHYSICAL SYSTEMS.

THE DESIGN PROCESS

Section V

The Nature of Design Education

20

Architectural Education

The hypothesis is that an effective design methodology has to encompass new design realities that need new procedures analogous to the operational principles that are meaningful in the natural and behavioural sciences. Furthermore education has to take up a different position toward these new realities which it has yet to explore. Since any procedure or observation will have to operate in the existing part of the world the conditions of this world will have to be clarified first before specifying any procedural framework.

The reality that the designer is addressing himself toward is no longer the safe one of easily applied principle and traditional model but the organization of four dimensional structures for multi-dimensional functions and activities over its lifetime; a very complex task indeed. The present simplicity of everyday procedures and tools with which this complexity is now being challenged leaves the profession and the schools quite incapable of coping with its dimensions. The professional means cannot even achieve relevancy let alone solutions to this problem.

On the one hand there are the approaches to reality formulated in the familiar terms of professional experience and the use of space and time terminology; the language and design apparatus to describe all phenomena of conventional environment and experience. On the other hand there is the multi-dimensional reality whose complexity exceeds these processes of description by magnitudes and with configurations that cannot possibly be described with such primitive tools. It is this misunderstood interface between the observer and the observed that has resulted in such gross distortions of the environment by the architectural and planning professions.

This general change of situation has most important results; there has to be an acceptance that deterministic concepts are definitely destroyed and probability and uncertainty introduced instead.

All ventures must aim now at a probable reality with uncertainty; in measuring one aspect the other must remain outside observation. All design efforts must be regarded as representations of a specifically probable world and that all efforts exert an influence upon each other which then changes the probability configurations. It is here that most academic institutions are reactionary in the extreme, when considering projects only of the highest probability, repeating only traditional experiences and losing all possibility and opportunity of influencing the future.

A necessary conclusion is that design must be understood as an inventive as well an systematic activity of projecting into an uncertain and only in probabilistic future, and not a rehash of the past. Deterministic statements are irrelevant, designers should work with the world not for it. Design means conceptualizing and anticipating the possible design behaviour by comparison of simulating models of real life over time, and the constant refinement of these models. Since it involves indeterminate forces and conditions, design can only precipitate a number of possible and therefore meaningful solutions; but not deterministic truth-cliches out of the rag-bag cupboard of educational or professional conventions.

This search for relevancy means leaving the cloisters of high academic and professionalism and getting involved in the on-going activities outside. The square manipulations and card-board cutting operations of todays schools belong to a different era altogether. The splendid isolation of today's students doing their own little programmes will have to disappear and approaches now outside the curriculum, as well as students from other diciplines, let in; allowing for inter-disiplinary study rather than occasional visits to other faculties. The connection should be an involvement in solving the problems in the environment; not training ficticious and reactionary 'Team-leaders."

The boundaries between academic education, the profession and the many activities of society should allow a cross-fertilization and feedback, which suggests continuous experiences with reality for all involved. Education should expand into non-academic learning and searching process, getting the feet and hands dirty; and conversely the reverse processes could alternate the educational establishment disturbing the cob-webs a little. Academic education might then be able to influence the external processes and have a chance to implement its own concepts and products which would be based on reality. But education would have to accept that its products should and must be marketable; which means using advertising mass-media techniques and the methods of scientific operation not strategies of academic heritage.

Much of the effort in today's education is still involved with learning skills instead of the development of sensory extensions and understandings, and the inventive experiences to break up pre-conceived set patterns. As a result old obsolete elements are replaced with new obsolete elements; still only using technology as an incidental aid. It remains the playing of the same old games with new rules in preference to new games altogether; not only the rules must be changed but the games as well.

Education will have to shift from instruction in out of date games and rules to exploration, discovery, probing and invention. The training of today's young people with yesterday's concepts, with yesterday's tools will have to stop, as will perpetuating yesterday's ideal by asking yesterday's questions. The new teacher must cease to be a prejudiced individual and become the inquirer into the total environment. An individual can only guide, instigate, and structure time to useful purpose while these investigations go on. He cannot usefully continue as a diseminator of unsubstantiated opinion but must be aware of the present and be able to understand its concepts and methods for problem solving.

It is clear that some very obvious changes will have to take place in architectural education before any progress can be made.

The facilities and equipment for design education need a revolutionary overhaul. The crowed studio, the drawing board, the T-square are still used as basic equipment for teaching and are mainly responsible for the actual structure of any course content. (It is analogous to teaching heart transplant surgery in a dark cupboard with pen knives with a mattress as a substitute for a live patient).

Any curriculum must remain flexible to provide a built-in corrective device to recognise obsolescence; it must be continuously evolving. To stay unchanging runs the risk of being left behind. The curriculum must be in the state of constant revision and reassessment and modified by experience so that it can achieve a desired result or thrown out if found wanting. The curriculum must be goal directed; those that operate it must know what they are doing and where they are going, which is in direct contradiction to present thoughts of letting everybody go where they want. Design education in the future cannot be viewed as merely manpower training for a shrinking profession. It will have to deal with the following propositions to remain viable;

It must look to interpretive analysis of current and future framework and trends. The emphasis must be on fresh insights and not after-the-fact wisdom; and on identifying an existing situation as it actually is so as not to have to rely on wishful thinking and looking in the 'rear view mirror.' To approach research, teaching and learning as an indivisible educational activity. Research should be an integral part of education, but research should not be made to mean the cataloguing of post-facto wisdom but a mission oriented data inventory for inventive concepts and constructs.

It must adopt the system approach to problem stating and problem solving. It must deal with hardware performance specifications and assemblange system design. Systems must not be confused with industrial components, although there must be an awareness of technological possibilities. The Architect must change from a poor sculptor to being a designer and builder of systems. Such systems cannot be constructed by individuals however well intentioned, but must learn to work in multi-disiplined teams with very diversified background and skills. To learn to work in teams is absolutely essential.

To become a builder of an industrialised system it is essentially to become involved in and to understand the manufacturing and assembly processes and the restriction placed on them by mechanical efficiency as well as the problems of codes, managerial problems, machine and tool use, erection difficulties, joining methods and all the other problems of the technological process in order to be able to reject the present classification and study of building types which so strongly structures solutions.

It must come to terms with the handle all economic factors of first cost and maintenance and running costs from the very beginning and not just tacked on at the end. For finance is the main and final arbiter in all discussions of buildings in the real world and is a measure of efficiency in problem solving.

It is essential that exploration takes place in new problem solving techniques by experimenting in new media and methodologies. Special experimentation projects should be undertaken and systems methodology made available to students so that they can be thoroughly familiar with its working. New thought processes, such as non-linear rationality, random theory, ambiguous pattern recognition, should be introduced. To over extend the mind can be as useful as to understand everything. The ability to structure problems is as important as the ability to solve them. All projects must emphasise the inter-relatedness and inter-connectedness between problems rather the individuality of them.

Ultimately there is a need in any educational faculty to give up the employment of overspecialised and similar thinking professional architects and change to a multi-disiplinary group whose connection is an interest in the total built environment. Inter-disiplinary work is inefficient and costly in time and money and near useless as a teaching method. Clearly structured mission oriented projects are a way of increasing efficiency as opposed to lashing around in the dark hoping for something to turn up.

Data is now increasing far too rapidly to be able to rely on traditional classifications. The alternative is that ludicrous behaviour caused by wrong classification will eventually lead to destruction. The ideas of fighting wars to preserve peace, smashing and defoliating a country so that its inhabitants might live in prosperty, slowly bankrupting and bringing a country to a standstill to save it from bankruptcy and stagnation, educating children not to think; are all actions that occur when decision makers and problem solvers have become completely dis-oriented about the relationship that links people together with their environment. The principles that are being insisted on, deny these relations and become less and less appropriate as it becomes more important to understand how the whole system works and less important to understand all about each little component.

There are immense problems to be solved that seem almost beyond solution because the catagory system does not seem to be able to deal with them. As with unpleasant moral problems there is now a great tendency to ignore the problems of expanding population, scientific knowledge, technological growth, knowledge growth, demands for increased education, and the destruction of nature and natural resources; and hope that by doing so that they will somehow vanish. It is a popular delusion that is these fields as in others such problems can be solved with a 'right policy.' But 'right' or 'wrong' policies are only another way of saying that one would represent an advantage of one group over another, it is nothing to do with the solution of the problem itself. The same applies to architectural policies for 'good' and 'bad' architecture or 'beautiful' or 'ugly' buildings; what is beautiful to one person may be ugly to another, neither truth nor beauty can be made manifest, and so which group is going to have its own way?

Classification attitudes, which are probably linguistic in origin, and which are emphasised by the educational system, suggest that all problems have to be approached two-dimensionally and not multi-dimensionally. Classification has up to now had preference over understanding; the society and therefore its environment and buildings have been looked on more like watches that can be taken to little separate pieces rather than like a jelly fish. Emphasis has been on taking things to bits and the more creative activity of putting things together in a different form for a different purpose has almost completely disappeared even from educational establishments. Synthesis, or the putting together of systems, is hardly understood at all even in places where such an understanding is central to the purpose of their existence. At best it is thought to be something that you ask children to do in order to keep them quiet.

It is quite possible that human thought and enquiry has reached a point where the taking of things to pieces has reached a stage that any future or further investigation is unlikely to produce many significant or useable results for an enormous expenditure of resources. It is probably no more than a useless exercise to work out exactly the form that a drop of oil on some water might take on under all circumstances although it could be done. Technological artifacts may have reached the same stage. Cars or refrigerators are unlikely to be superceded by anything radically different; not because something could not be thought up but because the society and culture is already too structured by them to accommodate such a radial change. To change them would mean to change the whole society. They are not separate components but part of a system and this is a mistake that technological and environmental prophets within the system and shifts have been in that position for some time, merely changing the size but more particularly their efficiency.

What is much more likely to happen is that these systems will become more efficiently linked together, their components will become more efficient and reliable and the understanding of the whole will allow choice about location to much greater advantage. It is possible that the aeroplane and the computer represent the last two major technological inventions to take place. Rockety is already showing that it has very little development potential or use as compared to its cost. Systems are going to get more complex and have an evolving system as a whole and to which everyone belongs and contributes.

The educational process which sets out to establish cultural norms, has always emphasised that Man is a rational being standing outside the environment, making it and destroying it much as he wished with little consequence to himself; and further of having an infinite capacity to understand what he was doing by taking it to pieces. The process was and is analogous to asking a person, who had no knowledge of the concept of time, to take a watch to pieces, and then say what its purpose was.

But the implication of Godel's Theorum are always with us. The human brain is constructed in a certain way and is therefore only capable of thinking in a certain way and so there are limits to what

195

can be completely rational thought and that which cannot be shown to be consistant and which might contain elements of chance and uncertainty.

The products of the human imagination should not be expected to fit tidily into the linguistic derived categories which we now possess and which are now being found wanting. These categories should be expected to change shape, to expand, to evolve. What is more the same form of answer to the same question cannot be expected. Answers to the question can come in different forms simply at the employment of different and multi-variable channels of communication; the Arts, Sciences, Mathematics, different cultures, different tongues, different classes, different income groups. A great deal of the problems in Architecture stem from the blind belief that drawings are the only way to state and solve architectural problems and if they cannot be drawn they cannot be of any concern to Architects.

It is very much open to doubt whether architects can now add anything new to the general aims, so far outlined, of environmental relationships. But the discovery and learning of these aims would open up an entirely new and hopeful prospect for environmental renewal on the basis of human evolution. Renewal relies on giving respect to and being allowed to, be guided by external biological and physical conditions.

In striving for such participation, the starting point should be awareness of the contemporary human situation; of particular relevance is the population increase; mechanised mass production of commodities; enhanced mobility; the state of knowledge about the guiding concepts and ideas of science and technology; what we really know about human beings and where they came from; their needs, physical and psychological; their behaviour; what happens when divergent cultures meet; and the world wide inter-relationship of human and environmental problems.

There is no doubt that there is an enormous difficulty in dealing with the achievement of a relationship between an awareness of this human situation and the environment. Such facts are entirely neutral and depend for their effect on human contact on as wide a basis as

196

possible; not on the mass dissemination of 'ideas' or 'policies'; but of people.

The increased possibilities offered by mass communication and mass mobility are already responsible for the breakdown of many national boundaries and could well help, in the long run, to alleviate the pressure of population in certain overcrowded areas, there being less and less reason to stay in them, their advantages becoming eroded by the possibilities of communication over much longer distances. Cultural objectives being just as transferable to tele-communications and person to person contact through greater personal mobility. Conversely there is not nearly so much need for general cultural agreement over large national areas because of the proximity of so many different view-points.

For the majority of educational establishments only the recent historic past and a history biased attitude count as design realities. In fact they compliment each other; the educational exposure of students to design problems that result most often from the professional and educational activities of the teachers generation and the student's future as a draftsman in offices, headed by representatives of the same generation and ideology; a clear craftsman approach to problems.

Within the context of this backward looking reality, academic and professional arguments and concepts compare well to the state of the technology and to the methods and tools in current use. The sophisticated backwardness of concepts with its wildly ambiguous terminology and operational plagarism matches the shiny technological antiquity of the so-called, almost Babylonian, building 'industry.'

Generally, any intelligent human enquiry searching for comprehension of complex systems underlying the organization of the material and living world involve the process of observing the real world and the conceptualizing and testing of possible models representing it; and in this way build up insight and knowledge of the objects and their behaviour, and to alter methods and concepts accordingly, not only as a result of external stimulus but also from reformulations of method.

Modern theories do not result from revolutionary ideas brought

197

from the outside, but their extraordinary importance is primarily established by a completely unexpected experience that arises in the normal procedures of observing the real world, and needs an explanation which is not currently available. New ideas arise from modification of traditional thought not abstract revolutionary concepts.

It is in this sense that educational institutions and the profession shows so little interest in searching for a comparable methodology of self-regulating feedback process between the accumulation and generation of information, and the construction and implementation of models that represent design realities. In this sense their activity can at no time really be classified as intelligent.

Neither the academic nor the professional groups use anything like adequate measuring tools for evaluating its design products preferring to start from scratch each time and ignoring the possibility of accumulated information. In spite of this operational helplessness, the problem does not just go away as they might hope. The reality of the culture and the society is constantly evaluating the built products of this wayward random design activity and constantly eliminating failures and permitting few monuments.

On the one hand, design activity means nothing until it is built in the environment where its interactions with social life determines its behaviour; behaviour which is a measure of its validity as a design; and which is at present outside the initial design process. And on the other hand this future behaviour is completely indeterminate initially and outside must become part of the overall concept, so that its information can be fed in and generate appropriate design criteria for the product's behaviour, over its life time.

Without this feedback, the design process now ends at a stage when the life of the designed object just begins and where all the following processes of the object and its interaction are external to the design. But even with vague feedback that does seem to exist now the flow of information is far too slow to be relevant.

Present design and construction activities operate on a time schedule that takes far too long a time interval between the initial decision making and the return of information from the performance of the product, to make it possible for any subsequent adjustments to be

198

made. This is further affected by the fact that life patterns are changing faster than the design and construction period.

Furthermore, the educational programs of the academic world are almost completely excluded from this process. Even the small amount of information that does get fed back concerning design and performance takes place in a world alien to academic education and is anyway a process that takes much longer than a student's academically available time.

21

Life is Right . . .

In the mechanical age, action and reaction were not closely connected in time and space, response was slow, involvement limited and consequences of actions remote and unreal.

In the electronic age and the rapid passing of information action and reaction are almost simultaneous; abolishing space and time.

Formerly the separation of action and reaction could mean non-involvement but now speed of communication and the passing of information, has created a world that Western Man is struggling in vain to avoid being involved in the consequences of his actions; and which look to an outsider as having all the appearances of the theater of the absurd.

One part of the culture, with its partial, specialized, individualistic, detached point of view, in its vindictive dying struggle to hold out a depth of awareness and unity that is becoming our time.

Images of this time are continuous, simultaneous, non-classifiable, non-codified, and run totally counter to the traditional compartmentalization of ideas and things, and counter to the analytical and rational processes of thought. Images are abstracted and require involvement and participation for their complete transference. They represent a continual flow of data measurable only in terms of the experiential information that they contain. The process of build up of many pieces of information giving the mosaic effect of composite impressions producing a fuller understanding.

The immediacy and amount of impression by the new communications systems applying to all the senses, making a fusion of them, and so giving the impression of being actually there and having to react; to being personally involved. The information coming through the electronic media is like an outgrowth of first hand experience and the immediate scene; not restated, redefined, edited, but real and sensed directly; no detached point of view whether a physical position or a state of mind is any longer possible.

For better or worse technological inventions have extended the body and now the mind, and so the environment; the situation has to be accepted and adapted to or the possibility has to be faced that we might be entering the age of the dinosaur rather than the age of unity.

A rigid deterministic world has given over to one of contingency and organic incompleteness and probability. Perfection and rationality are irrelevant; and architects oriented in this direction offer no interpretation or reconciliation with this time; only platitude and escape; preferring to derive their forms from the past in preference to the experiences of the present. First the present environment must be understood if it is to be controlled; its significance must be understood before structuring it and encapsulating it.

There is no doubt that search for new paths will carry on many of the old traditional attitudes to problem solving of the past, including those most firmly entrenched so that those who genuinely believe that they are setting out in new directions are in fact being misled by methods that are anything but new; simply adhering to traditional practices of solving problems with principles, images, carefully selected data and sheer expediency. The main dilemma is that having traditionally operated in the realms of Art and Architecture is having to find a new language that is more compatible with the methods of philosophy and science, but this means throwing away ideal precepts and visions that he is too loath to do; trying to incorporate them into some scientific sounding 'design method.' Trying to justify traditional practices for designing buildings with hopeful appeals to the methods of science, rather than throwing them all away and starting again.

The rational, analytical aspects of Architecture can be expected to give over to a non-classifiable accretion of elements in continuous uninterrupted flow without any particular sequence. As modern physics sees a universe in which everything happens precisely according to law, no longer, in which nothing is compact, tightly organised, or governed by strict, causality, so too will architectural impressions be less structured, ordered or controlled or in sequence. Impact is likely to derive from group effects, unselected, and unlimited, adjacent, oblique and marginal in experience.

As buildings become looser assemblages, less finite and static they will reach out and fuse their identity with adjoining constructions so that outside and in, yours and mine will appear so one continuous construction and a total environment where all the things connected with the human mind and experience can take place as opposed to the separation and monumentality of current architectural individualism in a past world of classified knowledge and exact definition.

If architectural elements disappear in this process they will not be replaced by new elements taken from the conventional sources but from wherever they may be found and made in much the same way that current art finds its sources and materials and techniques from wherever it needs to help express new ideas and concepts. The building industry as now constituted will give way to any sort of manufacturer who can supply the needed components and will fade away in the same way as the present architectural profession.

The 'cool' architecture of the future will be low in definition, in structure, and high in participation and significance. Architects and the occupants of these constructions cannot be detached or detach pieces for themselves in the sense that they will not take a detached or academic point of view, or one of personal isolation, because they are there; now.

There will no longer be any patience for the hypothetical, the make believe, the hypocritical, the isolated event out of context, with its sophistries, stunts, and mannered poses. Architecturally the historic revival, the literary reference, the moralizing, and the academic or fine-arts attitudes are dead. The environment can become a consumer commodity or an instrument for expanding and helping all to understand adjust to and enjoy the often bewildering environment of rapid technological change. It must be expected that the new architect armed with a new logic of architecture, aware of the vast growth and influence of the electronic revolution, with new perceptive attitudes and concepts, can become involved and participate in a consistant and valid expression of his own time of unity.

Selected Bibliography

ACKOFF, R. L. Fundamentals of Operational Research. John Wiley 1968.
ALEXANDER, C. City is not a Tree. Arch. Design.
  Notes on the Synthesis of Form. Harvard Univ. Press 1964.
  Systems Generating Systems. Arch. Design, Dec. 1968.
ARDREY, ROBERT. The Territorial Imperative. Collins, 1967.
  African Genesis. Fontana, 1967.
ASIMOW, ISAAC. Foundation
  The New Intelligent Man's Guide to Science. Nelson-Hall.
  Understanding Physics.
  View from a Height.
ASIMOW, M. Introduction to Design. Prentice-Hall.
ASHBY, ROSS W. Design for a Brain. John Wiley, N.Y. 1952.
BALDWIN, A. L. Theories of Child Development. John Wiley, 1967.
BAKER, C. A. Visual Capabilities in Space Environment. Pergamon Press 1965.
BANHAM, RAYNER. Theory and Design in the First Machine Age. Architectural Press London.
BARNETT, H. G. Innovation. The Basis of Cultural Change. McGraw-Hill 1953.
BELOFF, JOHN. The Existence of Mind. MacGibbon & Kee London 1962.
BERNE, ERIC. Games People Play. Grove Press 1967.
  Structure and Dynamics of Organisations and Groups. Grove Press 1963.
DeBROGLIE, LOUIS. The Revolution in Physics. New York, Noonday Press 1953.
BOGUSLAW, ROBERT. The New Utopians. Prentice-Hall, 1965.
BOULDING, KENNETH. The Meaning of the 20th Century. Harper & Row, 1964.
BROWN, J. A. C. Freud and the Post Freudians. Pelican 1961 and 1964.
BROADBENT, A. E. Perception and Communication. Pergamon.

205

BUNGE, MARIO. The Critical Approach to Science and Philosophy. Free Press of Glencoe 1964.

CALDER, NIGEL. Technopolis. MacGibbon & Kee 1969.
The Environment Game. Secker & Warburg 1967.
The World in 1984, vols. 1 & 2. Pelican ed. Calder.

CALHOUN, JOHN B. Population Density and Social Pathology. Scientific American Feb., 62.

CARSON, RACHEL. Silent Spring. Fawcett Public. Greenwich, Conn. 1962.

CHAPANIS, A. Man Machine Engineering. Johns Hopkins. 1959.

CHARTER, S. P. Man on Earth. Contact Ed. Sausalito, Calif. 1962.

CHERRY, C. On Human Communication. Science Eds. Inc., N. Y. 1963.

CHURCHMAN, C. WEST. The Systems Approach. Delta, N. Y. 1968.

COMMONER, BARRY. Science and Survival. Viking Press, N. Y. 1963.

DE BELL. The Environmental Handbook. Ballentine Books, N. Y., 1970.

DE BONO, E. The Uses of Lateral Thinking. Allen Lane.
5-Day Course in Thinking. Basic Books.

EDWARDS, E. Information Transmission. Chapman Hall 1964.

EHRLICH, PAUL. The Population Bomb. Ballantine Books, N. Y. 1968.

EIBL EIBERSFELDT, I. Flight Behaviour of Animals. Scientific American Dec. 1961.

EMERY, FE., ed. Systems Thinking. Penguin Mod Management 1969.

FOSS, BRIAN. New Horizons in Psychology. Penguin Original 1966.

FREUD, SIGMUND. The Interpretation of Dreams. Basic Books 1965.
Two Short Accounts of Psycho-Analysis. Pelican 1962.
Beyond the Pleasure Principle. Bantam Books.

GABOR, DENNIS. Inventing the Future. Pelican 1964.

GAGNE, R. ed. Psychological Principles in Systems. Holt, Reinhardt & Winston 1962.

GALBRAITH, J. K. The Liberal Hour. Penguin.

GAMOW, GEORGE. One-Two-Three Infinity. Viking 1961.
Biography of the Earth.
The Birth and Death of the Sun. Mentor.

GARDNER, M. Ambidextrous Universe. Allan Lane Penguin Books.

GEORGE, F. H. Cybernetics and Biology. Oliver and Boyd 1965.

GERADIN, C. Bionics. World University Lib. 1968.

GRAVES, ROBERT. The White Goddess. Farrar Strauss 1966.
King Jesus. Cassell.
The Greek Myths 1 & 2. Penguin.

GREGORY, R. L. Eye and Brain. World University Lib. 1966.

GORDON, W. J. J. Synetics. Harper & Row, N. Y. 1961.

HOYLE, FRED. The Nature of the Universe. Signet.
Frontiers of Astronomy. Heineman 1965 (ob. Murray Books).
Galaxies, Nuclei and Quasars. Heineman 1965.
Astronomy. McDonald 1962.

HEISENBERG, WERNER. The Physicist's Conception of Nature. London, Hutchinson 1958.

HELMER, D., & T. GORDON. Social Technology. Basic Books, N. Y. 1967.

HERBERT, L. pp. Present State of Design Method. Architect '69 1968.
Villages in Northern Ghana (in assoc. Prussin) Photography, Drawings. U.C.L.A. Press 1969.
Extracts. Art & Architecture 1967. World Architecture 1967.

HOFFER, ERIC. The Ordeal of Change. Harper & Row N. Y. 1963.
The True Believer.

HUTCHINGS, ed. Frontiers of Science-ed. essays. Allen & Unwin 1960.

JACOBS, JANE. Death and Life in Great American Cities. Random House, 1961.

JONES, J. C. & D. G. THORNLEY. Conference on Design Method. Pergamon.

KAUFMANN, ARNOLD. The Science of Decision-Making. World University Lib. 1968.

KENNISTON, KENNETH. The Uncommitted. Delta 1960.

KOESTLER, ARTHUR. Insight and Outlook. Macmillan.
The Act of Creation. Hutchinson 1964.
The Ghost in the Machine. Hutchinson 1967.
The Sleepwalkers. Hutchinson 1959.

KUHN, THOMAS, S. Structure of Scientific Revolutions. University of Chicago 1962.

LORENZ, K. On Aggression. Methuen.

LEVI-STRAUSS, C. The Savage Mind.
Structural Anthropology. Basic Books 1963.

McFARLANE-SMITH. Spatial Ability. U. of London Press.

MASTERS, R. E. L. & J. HOUSTON. Psychedelic Experience. Anthony B10nd Ltd. 1967.

McLUHAN, MARSHALL. The Gutenberg Galaxy. Routledge & Keegan Paul London 1962, rev. 1967.
Understanding Media. Signet 1966.
Article in "Playboy." March 1969.

MEDAWAR, P. B. Induction & Intuition in Scientific Thought. Methuen 1969.

MILLER, G. A. Language and Communication. McGraw Hill 1961.

MILLER, GALANTER, PRIBAM. Plans & The Structure of Behavior. Holt, Reinh. and Winston.

MORRIS, DESMOND. The Naked Ape. Jonathan Cape.
The Human Zoo. McGraw Hill 1969.

MOHOLY-NAGY, L. The New Vision. Wittenborn, Schultz 1949.

MONTAGU, ASHLEY. The Science of Man. Odyssey Press, N. Y. 1964.

MORONEY, M. J. Facts from Figures. Pelican.

MULLER, PHILLIPPE. The Tasks of Childhood. World Univ. Lib. 1969.

207

N.A.S.A. Bio-Astronautics Data Book. U.S. Gov. Print. Off. Washington, D. C.

NICOL, HUGH. The Unity of Man. Constable 1967.

PADDOCK, W. & P. Famine 1975. Little, Brown 1967.

Papers and Proceedings.

P.A.R.S.E.C.

HERBERT, L. The Meaning of the Word Design. Oct. 1967.
Possibilities—The Future of the Architect.

YEOMANS. D. Matters of Fact and Matters of Opinion.

MADDOCKS, J. Education.

HERBERT, L. Language Problems in Architecture Jan. 1968.

YEOMANS, D. Thought-Design Process.

HERBERT, L./YEOMANS, D. Report of Education Committee 1967.
Oxford School of Architecture.

PIAGET, JEAN. The Origin of Intelligence in the Child. P. B. Norton 1963.
The Child's Conception of the World. P. B. Littlefield 1960.
The Child's Conception of Physical Causality. P. B. Littlefield 1960.
The Child's Conception of Number. Humanities.
Judgement and Reasoning in the Child. Humanities 1947.
The Moral Judgement of the Child. P. B. Free Press.
The Language and Thought of the Child. Humanities 1959.
The Psychology of Intelligence. P. B. Littlfield 1960.

POLYANI, MICHAEL. The Study of Man. Kegan. Paul, London 1959.

POPPER, SIR KARL. Conjectures and Refutations. Basic Books 1963.
The Logic of Scientific Discovery. Basic Books 1959.

PROGRESSIVE ARCHITECTURE. Dec. 1966. May 1969.

PYKE, MAGNUS. The Boundaries of Science. Pelican.

RAPOPORT, A. Fights, Games and Debates. Univ. Michigan 1960.

REINOW, ROBERT & LEONA. Moment in the Sun. Ballantine, 1967.

RIODELLE, A. I., ed. Animal Problem Solving. Penguin Books 1967.

RIVETT/ACKOFF. A Manager's Guide to Operational Research. J. Wiley & Son 1963.

SAWYER, W. W. Mathematician Delight. Pelican-Penguin 1943.
Prelude to Mathematics. Pelican 1955.

SCIENTIFIC AMERICAN. Sept. 1966 Information/Computers.
Sept. 1965 Cities.

SCIENCE JOURNAL. Forecasting the Future, Oct. 1967.

SHAFFER, RITTER, MEYER. The Critical Path Method. McGraw-Hill 1965.

SNOW, LORD CHARLES PERCY. The Two Cultures and the Scientific Revolution. Cambridge University Press 1969.

SMITH, JOHN MAYNARD. The Theory of Evolution. Pelican, 2nd ed. 1966.

SOMMER, ROBERT. Personal Space. Prentice-Hall Spectrum PB 1969.

SPROTT, W. J. M. Human-Groups. Pelican 1958.
STEARN, G. E. ed. McLuhan Hot and Cool. Penguin 1968.
TRICKER, A. A. R. The Contribution of Science to Education. Mills and Boom 1967.
The Assessment of Scientific Speculation.
VERNON, M. D. The Psychology of Perception. Penguin 1962.
Experiments in Visual Perception. Penguin 1966.
WALTER, GREY W. The Living Brain. Norton, N. Y. 1953.
WIERER, NORBERT. The Human Use of Human Beings. Sphere Books, London 1968.
Cybernetics. John Wiley N. Y. 1948.
WHITHEAD, ALFRED NORTH. Adventures of Ideas. MacMillan 1933.
WILSON. Information Communication of Systems Design. John Wiley 1965.
WITTKOWER, RUDOLF. Architectural Principles in the Age of Humanism. Tiranti, London.
YOUNG, J. Z. Doubt and Certainty in Science. Oxford Univ. Press 1961.

Appendix

Three Basic Concepts Used In This Book

GODEL'S THEORUM

GODEL'S THEORUM points out that in order to talk about a strictly mathematical system of logic in fact a meta-language must be used which in itself is not logical at that moment in time. As it is explored and a part reached which is made consistant, then there is another level which is always incapable of analysis in these terms. To advance, the frame of reference can and must be altered because whatever the completeness of an experience there is always a further step which denies complete understanding.

HEISENBERG'S PRINCIPLE OF UNCERTAINTY

Heisenberg's Principle of Uncertainty, which outlines the unobservable property of nature, says specifically that it is impossible to determine simultaneously both the position and momentum of a particle. In effect, it says that no one can know where they are going unless they know where they are and how fast they are going in any direction at any given moment. As a result of this PRINCIPLE causality is seen as indeterminable; 'x' does not cause 'y', it simply preceeds it; or happens simultaneously. The concept of relative probability, based on the probable outcome of a large number of events has to take the place of the concept of cause and effect. A chain of events cannot take place.

213

It cannot, however, be assumed that an event that has taken place in the past will take place again in the future in anything except sub-atomic particles, as the numbers are not nearly great enough even for a probability. Historic events are not only unlikely to take place again but the observer's act of observation is likely to change the evidence which is a further concept to arise from the PRINCIPLE OF UNCERTAINTY.

The nature of matter cannot be known except mathematically. An equation can be written to describe an atom's behaviour but a scale model can never be made of it since it is not a rigid structure of particles but one of energy states and resonances. The probable position of electron around an atomic nucleus can be predicted mathematically; for instance the probability of them being inside the nucleus is almost nil; but their exact location is unknown, as is the shape they take up around the nucleus.

Events are such that scale models fixing them in one position at one time can never be made in a way that they correspond to reality; they have to be considered as constantly unfolding events in time not as static entities suddenly changing. The process whereby this happens is described by evolution and its mechanisms.

EINSTEIN'S THEORIES OF RELATIVITY

Einstein's Theories of Relativity altered not only traditional modes of perception, but also completely changed the meaning of four abstract concepts basic to the conception of the world; 'space,' 'time,' 'matter,' and 'motion.'

SPACE, TIME, MATTER, MOTION

To most people 'space' is defined as an invisible air that fills the gaps between 'objects' and all objects exist in space like pinpoints or dots on a clear background. Einstein undermined this notion because it is immeasurable and useless as a definition and replaced it with the concept of relativity which says that 'space' is any reference frame chosen at any moment that it is wanted. To clarify the idea of a reference frame it must be imagined that there is a man on a train with his suitcase on the rack above his head. It appears that the suitcase is at rest relative to the passenger; it does not fall off the shelf until the train stops suddenly; then it falls on his head. Was it at rest relative to the train and in motion relative to the immovable earth or was the earth in movement relative to the train and the suitcase? For a man outside the train the earth was at rest, for the man inside the earth moved.

For most events on earth the earth itself is used as a fixed frame of reference to measure whether other things are moving; and this is what is understood by 'motion.' But it is not an absolute, to see whether anything moves there has to be something that does not move, and that something has never been found even at the farthest ends of the universe. Everything is moving relative to everything to everything else but at different speeds or in different directions. The point is best illustrated by imagining a cowbody film in which the hero leaping from his horse onto a run-away stage coach containing the beautiful heroine in order to save her from her fate. He rides his horse alongside and leaps aboard in a seemingly suicidal man-oeuevre, but a closer look will show that the coach and horse were traveling in the same direction and at the same speed which means relative to each other they were at a standstill and the operation no more difficult than stepping inside a front door of a house.

Absolute 'time' is the classic way to see 'time' as a relentless, uniformly, irreversible forward movement like a current. But relativity has shown that time cannot be measured apart from events and must therefore be regarded as another dimension that shapes and determines events, not an absolute or abstract frame of reference for events to take place. But when an event stops so does 'time' for it. Only the clock keeps going because there are many events in a day which is itself an event. Most people are aware of the effect of 'speeding up' time relative to the clock when they are interested in an event and the opposite effect when they are bored.

Essentially what relativity has done to abolish 'space' and 'time' because they cannot be measured independently of particular events in which measurements differ from observer to observer. As the frame of reference is changed so are the events and the way they are seen.

215